CW01582178

Field Guide to the Wildflowers of Australia's South West

Augusta — Margaret River Region

Jane Scott
Illustrations by Patricia Negus

Cape to Cape Publishing

First published in September 2002 by
Cape to Cape Publishing
North Fremantle, Western Australia 6159

Copyright © Jane Scott, and Patricia Negus 2002

This book is copyright. Apart from any fair dealing for the purpose of private study, research, criticism or review, as permitted under the *Copyright Act 1968*, no part may be reproduced by any process without first getting written permission from the publishers. Requests and inquiries concerning reproduction and rights should be made to the Manager, Cape to Cape Publishing, 50 Harvest Road, North Fremantle WA 6159.

Cataloguing-in-Publication entry:

Jane Scott

Field Guide to the Wildflowers of Australia's South West
Augusta – Margaret River Region / by Jane Scott ; with illustrations by Patricia Negus. –
North Fremantle, W.A. : Cape to Cape Publishing, 2002.
224 p. ; ill. ; 30 cm.

Includes index.

Hardback: ISBN 0 9577729 1 2
Paperback: ISBN 0 9577729 2 0

I. Negus, Patricia. II. Title
1. Wildflowers - Western Australia - Augusta-Margaret River Region - Identification
2. Wildflowers - Western Australia - Pictorial Works

Photos: Jane Scott
Produced and Printed entirely in Australia.
Cape to Cape Publishing,
50 Harvest Road, North Fremantle WA 6159, Australia
Production & typesetting: Ray Forma
Typeset in 9pt Times and Helvetica
Scanning & film separations: Show-Ads Digital Centre, West Perth WA 6005
Printing: PK Print, Hamilton Hill WA 6163

In memory of Georgiana Molloy

Georgiana Molloy, of Augusta, was passionate about, and meticulously documented the flora of this far south-west corner of Australia during the early years of British settlement. In June, 1840, she wrote in a letter to English horticulturalist, Captain James Mangles:

> "such flowers of imagination; I am now in raptures when I think on them; in searching to come suddenly on such gems and be surrounded by them, makes you for a time think you are in fairyland".

Quotation from letter sourced from:
Lines, William J. 1994. *An All Consuming Passion: Origins, Modernity, and the Australian Life of Georgiana Molloy*, Allen & Unwin.

Contents

Opposite: ***Banksia grandis***, Bull Banksia (described page 58)

P.Negus

Foreword

The south-west corner of Australia is a virtual island of plant species richness. Within this region there are special areas of biological and landscape interest and one of these is the well-watered far south-west corner, around the now famous Margaret River district.

The region has tall forests, woodlands, extensive areas of swamp vegetation and an incredible array of soil types. Sandy soils abound and so too do rich loams and lateritic uplands, each with their own special vegetation cover. The coastline is spectacular and in its undisturbed state is clothed by a huge variety of native plants able to withstand the winds and salt spray. A major problem confronting the local and the wider community is how to make this flora more widely appreciated and to manage and preserve it for future generations.

A book such as this has a key role to play in conservation. Knowledge about the plants of any area needs to be made readily accessible to the general community, locals and visitors alike. A community that is knowledgeable about its local flora and values its uniqueness can promote its conservation.

The authors of this book deserve high praise for presenting the reader with wonderful images of their rich regional flora with their correct Latin names. They and their colleagues are active members of the WA Regional Herbarium Project; a scheme that assists regional groups like theirs to access up-to-date information about the WA flora. The authors have successfully welded their vast local experience and enthusiasm with that of the States botanist's knowledge of names to produce this incredible book.

Neville Marchant
Herbarium
Department of Conservation and Land Management

Acknowledgments

The production of this book has gone hand in hand with the Community Regional Herbarium Volunteer Program, an initiative of the Western Australian Herbarium in Perth. We are most grateful to all the WA Herbarium staff and volunteers involved for their work in identifying and processing our specimens and for their continuing encouragement and enthusiasm and in particular to Jan Gathe and Mike Hislop for their valuable help in editing my rather erratic text. Thanks are also due to Judy Wheeler (Department of Conservation and Land Management, Albany) for her comments on *Agonis* and *Hibbertia*.

We thank the WA Herbarium for giving access to their *FloraBase* web-site, which provided invaluable data, especially on plant distributions.

Thanks also to those involved with the organisation of the Regional Herbarium in Margaret River, especially to Jo Alferink for her efforts in mounting and displaying at the Margaret River library the plant collection that compliments this book.

Brian Hanich, from the WA Museum, was most helpful with information on galls and their inhabitants.

We sincerely thank our many, many friends who have shown so much interest and support for our project, in particular Margaret Leavesley, whose vigilant observation led to some interesting finds on her delightful bush property; also to Molly Hall, Arthur Budge and Jenny Stevens for valuable snippets of information and for collections on their properties.

We are grateful to CALM, the Department of Conservation and Land Management, for permission to collect in reserves and crown land for the Regional Herbarium.

We have drawn on many references (listed in the Bibliography) for information contained in the text and, in particular, wish to acknowledge F.A. Sharr's publication, *Western Australian Plant Names and their Meanings*, which was the source for meanings of the scientific names, and *Western Weeds* by B.M.J. Hussey et al. for valuable information on the weeds.

Not least, thanks are due to our patient families who sometimes took a back seat as we became totally engrossed in our mission!

Introduction

The flora of south-western Australia is incredibly rich. Over 8,500 native plant species and many more sub-species and varieties have so far been named in Western Australia, most in the south-west, and more and more are being identified as our knowledge expands. Inevitably, general books on the wildflowers of the region can only cover a very small proportion of this amazing diversity. Even armed with half a dozen of the very good publications available most of us will be lucky to give species names to more than a tiny fraction of the plants that we find on a spring walk. Detailed up-to-date references that are needed to confirm identifications, are just not available in many genera and there are question-marks on many species, even in this well-known corner of the state.

With this in mind, the *Field Guide to the Wildflowers of Australia's South West* attempts to document as many as possible of the wildflowers of our home area. By doing this, we hope to give friends, residents and visitors to this beautiful part of the south-west an opportunity to become more familiar with the flora of this very special part of Western Australia.

We aim to show the mind-boggling array of plants that grow within the limited boundaries that we set ourselves, to highlight their intricate beauty and to give some understanding of the relationships that exist between them. As well as the more familiar wildflowers that may be found around the area, we also include many of the unobtrusive little plants that grow here. Some of these many people may not even be aware of, while some are often seen in the bush, but are rarely featured in wildflower books. Many of the plants that grow here are widespread throughout the south-west corner of the state and we hope that this book will be a useful addition to the reference library of anyone who wants to learn more about this region's fascinating flora.

The Community Regional Herbarium Volunteer Program

It was coincidental that we decided to begin this project when local communities were being encouraged to participate in the Regional Herbarium Project. This provided a focus for our efforts and the support and information that we received from the WA Herbarium helped us enormously.

The Regional Herbarium Program encouraged volunteers from all over the state to make plant collections from around their homes. This increased participants' knowledge of local plants whilst, at the same time, providing a wealth of specimens to add to the State Herbarium. In this way the project contributed to an improved knowledge of the state's floral diversity with the discovery of several new species and many range extensions, including some for plants considered rare or poorly known.

The regional volunteer collected two specimens from a particular plant or site, labelled and pressed them, and made notes about the plant, its habitat and other species growing there. One specimen was retained for the local collection, while the other was sent to the State Herbarium to be correctly named and entered into a database. The State Herbarium then sent the correct name and numerical barcode of each specimen, together with informative notes, back to the Regional Herbarium. The State Herbarium also updated the Regional Herbarium specimen with any name changes.

Many of the species illustrated in this book, with the exception of the orchids and a few others, have been painted from specimens 'vouchered' in this way. For each of these, we have included the WA Herbarium (**PERTH**) identification barcode number; for example ***Kingia australis* 05612004**. Thus you can check them later for name changes against the corresponding specimen held in the Margaret River Regional Herbarium, or with its equivalent in the WA Herbarium in Perth. Our Regional Herbarium is currently housed in the Margaret River town library. Unfortunately, as we go to press, the future of this valuable program is in doubt, due to lack of funding.

Drosera menziesii subsp. ***menziesii***, Pink Rainbow (described page 88)

Scope of this book

All the plants in this book are found within the western half of the Augusta–Margaret River Shire. The Shire boundary was chosen as our rather arbitrary northern limit and we collected plants from the coast eastwards approximately to a line running north from Hardy Inlet, 5–6 km east of Bussell Highway. As well as the coast and the Leeuwin–Naturaliste Ridge, this area includes a substantial part of the Jarrah forest without venturing too far inland, where the mix of species changes significantly.

These spatial boundaries covered an area that we considered manageable. We also gave ourselves a three-year time limit. We envisaged that the project could otherwise stretch well into the future, as we discovered 'just one more' species to include! Even so, many plant groups are represented by only a small sample of the species that grow here. There are probably over 100 orchids, for example, that grow within the area, but these are covered well in more detailed publications. We know that other large groups, for example, trigger plants (*Stylidium*) and sundews (*Drosera*), are represented in the area by more species than we have illustrated. We have also attempted only a very small sample of some of the more easily identified rushes, sedges and grasses. We have much more to learn about these plant groups and maybe a further volume will cover them in a few years' time! But, with these important exceptions we hope to have included most of the indigenous plants and some of the more noticeable weeds that grow in the area. However, production of a book such as this only serves as a starting point from which our knowledge can expand and we welcome feedback on any plants that we have missed, extensions of range or interesting details that we have not recorded.

Arrangement

The book contains entries for 500 plants. The indigenous species are arranged taxonomically, that is according to their **family**, and within this by their **genus**. In this way related species are grouped together. Most introduced (weed) species have been treated in a separate section, although some have crept into the main part of the book. Once you develop an 'eye' for the look of the main plant types, it should be fairly easy to find the appropriate section for the plant that you are trying to identify. Of course there are many, less obvious species that will require more searching, but this should become easier as you become familiar with the layout of the book.

Although we had mapped out the general arrangement of species from the outset, it was inevitable that we would discover some new and some forgotten plants during the course of the project. Some we were still able to fit into the space available. Where this was not possible, they have 'spilt over' onto the text pages and, in one or two cases, may be slightly out of sequence. Such is the nature of a dynamic project!

We felt it was also very important to include the introduced plant species that are most commonly encountered in the bush. As stated above, they are mostly treated separately, but some have been included in the main part of the book for easy comparison with closely-related native species, or where they have become naturalised here over a long time period. All aliens are marked with an asterisk * in the text.

After the illustrations, a section of the book briefly describes the area's main habitat types and we have provided a check-list of species that you can use, for example, to record the plants found on your property, around your home, on particular walks in the bush, or for those flowering in different seasons.

Species descriptions

One of the features that we are often asked about is the meaning of plant names. Where possible therefore, we have included a brief meaning of the plant's scientific name and, in some cases, also the common name. We hope you will enjoy brushing up on your Greek (G.) and Latin (L.)! Many of our plant names also have interesting historical connections with early botanists, plant collectors and other prominent people. In many instances, once you know the meaning of a name, it becomes much easier to remember. We are indebted to the excellent publication: *Western Australian Plant Names and their Meanings* by F.A. Sharr (UWA Press) that was our source for this information.

Drosera sulphurea
(described page 88)

The plant descriptions are brief and aim to compliment the paintings by giving details of the plant's form, shape and size, the type of habitat in which it grows and general notes on its status and distribution. We also indicate the approximate flowering time **for our area**. We have included particular points of interest about some plants, but have not attempted to describe plant uses in any detail. Another whole volume is waiting to be written on this subject.

Priority Flora

Some of the plants in this volume appear on the Priority Flora List of CALM and are marked with a △. These may include flora that are rare or threatened, but for which there are insufficient data to accurately determine their status, or those that are considered rare, but not currently threatened. Priority Flora do not have the same legal status, and therefore protection, as Declared Rare Flora, but the Priority Flora List is maintained to highlight flora of special conservation interest and to encourage their appropriate management.

Plant diversity in Australia's south-west

The flora of the south-west of Western Australia is one of the most unique and diverse in the world. The flora of south-west WA today, has developed from an ancient, common flora that originated on the vast super-continent of Gondwanaland. This then evolved into distinct but related plant groups, as this land split up into the modern continents of South America, South Africa, India, Antarctica and Australia.

While south-west Australia has been separated from the rest of Gondwanaland for only a short time, geologically speaking, the landmass itself is extremely ancient and has been very stable for millions of years. Its soils are therefore very old and are extremely poor in nutrients because there has been little 'renewing' volcanic or glacial activity for a very long time. Much of the south-west has remained above sea-level for over 200 million years. This stability has given time for a tremendous diversification of plant species, each evolving to fill a different niche and each developing a particular strategy to cope with the wide array of impoverished soil types. Changing climatic regimes over this long period have added to the variation in soil types that, in turn, has encouraged further diversification amongst the plants.

The positioning of the south-west has made it a biological 'island', cut off from other plant communities by oceans to the west and south and by deserts to the north and east. Thus there has been very little intermingling with other floras for a very long time. As a result, many endemic species, unique to the south-west have evolved. On a smaller scale patches of different soil types also act as islands, each hosting its own complex of species, some of which have only very localised distributions.

The long-term stability of the region has also led to the development of the myriad of intricate and subtle relationships between plants and the mammals, birds and especially insects that pollinate them. This has further added to the huge variety of our flora.

The greatest diversity in the WA flora is found in the Kwongan, or sandplain heathland plant community, that is mainly located to the east and north of the south-west forests. However, the Margaret River region is still home to a surprising number of plant species. As well as the coastal, heathland and woodland habitats of the Leeuwin–Naturaliste Ridge, the area lies at the meeting point between Jarrah–Marri and Karri forest types and many species therefore exist at the northern or southern limits of their range. Jarrah–Marri forest, especially, hosts a huge range of understorey plants. Quite distinct plant communities may often be found within a few metres of each other, each dependent on quite small differences in soil type.

When and where to look for wildflowers

The Augusta–Margaret River area is very fortunate in still retaining a substantial amount of beautiful bushland. The Leeuwin–Naturaliste National Park stretches all down the coastal ridge to Cape Leeuwin near Augusta and also widens to include the lovely Boranup Karri forest, about 20 km south of Margaret River.

There are also other small reserves west of Bussell Highway and large areas of State Forest to the east. Most of these blocks can be accessed easily by car and finding somewhere where you can safely pull off the road, park and go for a potter in the bush is a relaxing and rewarding pastime in spring and early summer. There is always something to find, but in September, October and November the flowers are at their most prolific and the mix of species that are flowering changes from week to week. You do need to get out of your car and go looking! Many people come down from the north of the State, having driven through spectacular fields of everlastings and wonder why they see nothing from their car windows in the south-west heathlands and forests.

Roadside reserves can also be very rewarding, even if they are fairly narrow. The road out to Gracetown from Caves Road, the small subdivision roads along the Leeuwin–Naturaliste Ridge near Prevelly Park and around Augusta, for example, have lots to offer. Once again, make sure you can stop safely and pull well off the road.

Many of the area's tourist attractions also have lovely bush settings and are very picturesque in spring, as are some of the driveways to vineyards and wineries.

If you are walking or riding a bike there are many wildflower opportunities, such as the bike track from Margaret River to Prevelly, or the old railway reserve between Margaret River and Cowaramup. The track from town along the north side of the Margaret River passes through superb bushland right out to the Ten Mile Brook water-supply dam. From Augusta you can head north along Hardy Inlet onto the fascinating granite outcrops. Along the coast there is the Cape to Cape Walk Track and there are many other walking options within the National Park. The car-parks and short walk tracks around the public caves are excellent venues in which to fossick for wildflowers and so are the National Park campgrounds.

Note. For very sound reasons it is illegal to pick wildflowers without a relevant licence. Most do not last well in any case, so please respect this law. We are lucky to have many large and colourful flowers that are easy to photograph, thus making a much more lasting record.

When is a weed not a weed?

Indigenous, or native, species are those that have evolved here over time, developing complex relationships with the other plants and animals within the local environment. Together, these organisms form communities that are ever-changing and dynamic, but are basically in balance. When a species is introduced from another country, or even from another district, it does not bring along its associated pests, diseases or predators, or even the limiting environmental factors that kept it in check within its home range. It may therefore run rampant, crowding out and choking the native vegetation that cannot compete with the alien's new-found vigour. *Leptospermum laevigatum,* (see p. 196) is an Australian plant, but is native to Victoria. It was introduced to the south-west as a garden plant, but has run out of control in many places, such as at Ellensbrook.

Interestingly, while many of the weed species in the south-west have been introduced from South Africa, some of our own West Australian species have been transferred in the other direction. Some of our *Acacia* species, for example, were planted for use as nutritional fodder, but spread out of control and have now become weeds in South Africa.

It does not necessarily follow that an introduced species will become a nuisance here. There are many aliens that simply find a niche in the environment, without threatening their neighbours. Some of our coastal plants especially, were introduced long ago and have found their place in the ecosystem. Indeed, in some cases they have become most useful colonisers and stabilisers of mobile dune systems. However, weeds are a huge problem everywhere, threatening the viability of our unique flora. Finding the means of eradicating them effectively is very difficult and costly. Developing our awareness and recognition of weeds is the first step in this direction.

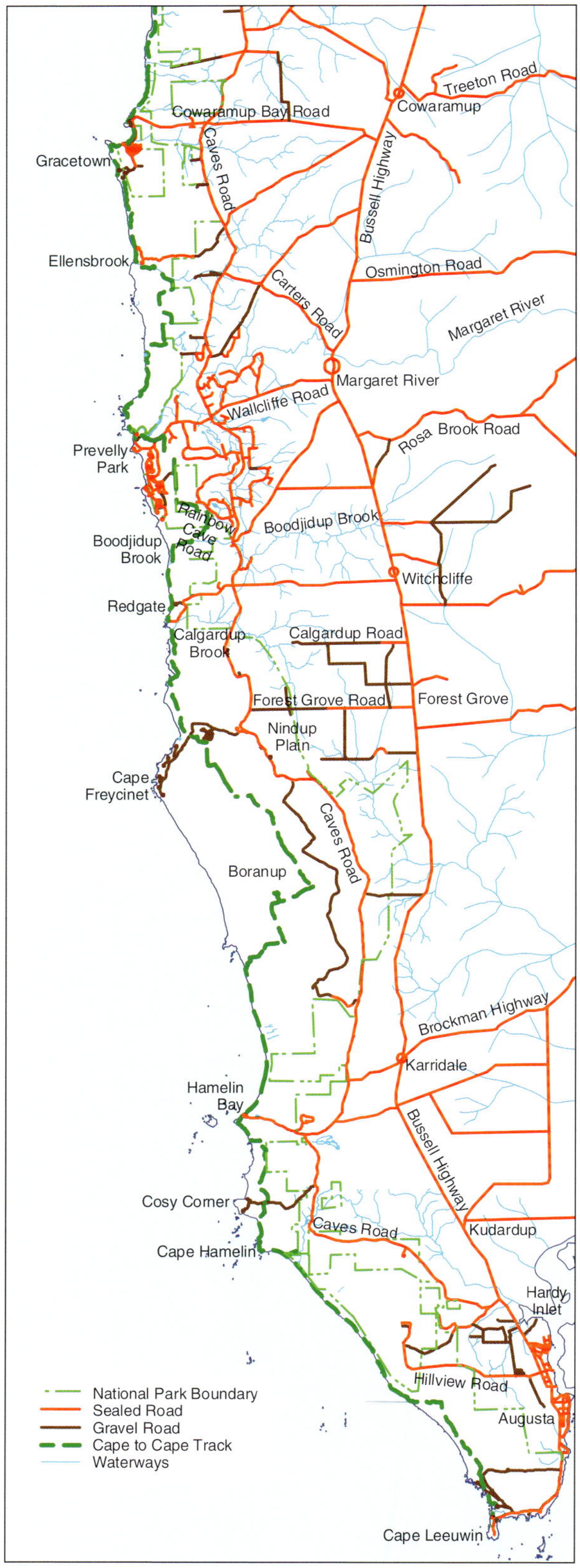

Location Map of the collection area – The Western half of the Shire of Augusta – Margaret River (1:300,000)

The paintings by Patricia Negus

This has been a wonderful project for me. I have learnt so much about our local flora and, after studying so many of our plants at close quarters, I am simply amazed at the very many ways their parts are put together, giving us such diversity amongst our flowering plants. I hope that you, like me, will get to know these plants and their names, so that when you come upon them on a bush walk, it will be like meeting old friends.

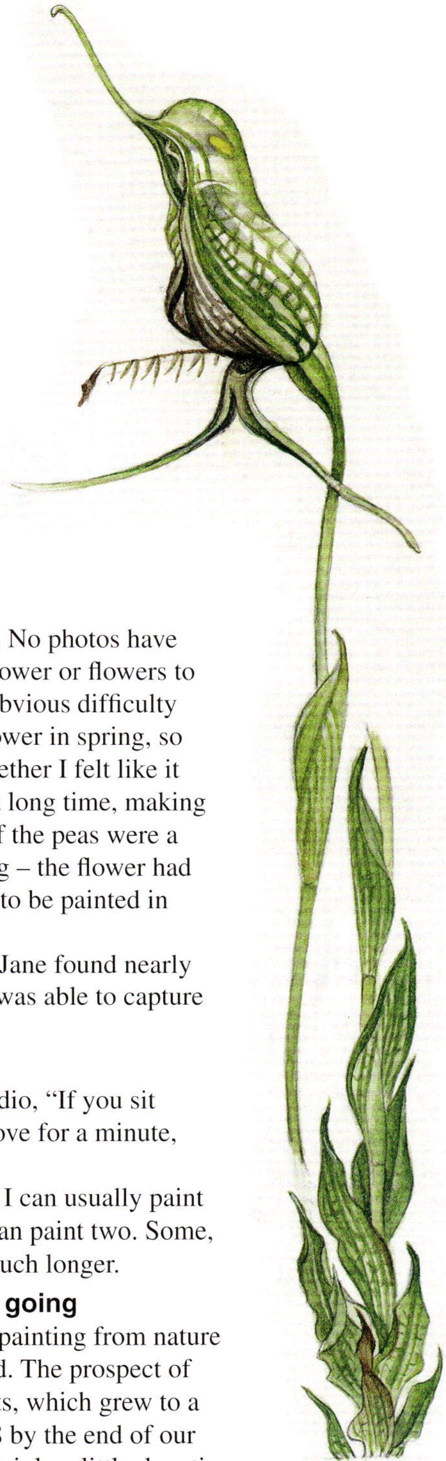

Why paintings and not photographs?

We all know how wondrously beautiful photographs can be. But the advantage of a painting is that all the parts of the plant that are needed for its identification: buds, flowers, fruits, leaves, etc. can be brought into focus, and arranged together. Painting also enables related plants to be grouped together on a page, and intertwined with one another. I have also included many enlargements of small parts of the plant that I have found interesting, but which cannot be seen clearly with the naked eye. You will see them too, if you make a habit of carrying a pocket lens on your rambles.

The difficulties of painting wildflowers

Every illustration in this book has been painted from actual specimens. No photos have been used. Jane would usually go out early in the morning and bring the flower or flowers to my studio, where I would immediately start to draw and paint. The most obvious difficulty is that plants don't flower at regular intervals throughout the year. Most flower in spring, so September, October and November found me painting day in, day out, whether I felt like it or not. Some, like the trigger plants and the orchids, would stay fresh for a long time, making them a delight to paint. Others, like the flags, wilted very quickly. Many of the peas were a problem; one minute I would be happily painting away, and the next – ping – the flower had closed! Some of the sundews and fringed lilies wouldn't last at all, so had to be painted in the bush.

It was a real challenge of timing for us both. It was quite amazing that Jane found nearly every plant that she was looking for, and many more, in flower, and that I was able to capture them all in paint.

How long does it take to paint a flower?

My most asked question! It reminds me of something I heard on the radio, "If you sit with a pretty woman for an hour, it seems like five minutes; sit on a hot stove for a minute, and it seems like an hour!"

For me, the whole day passes and it seems like five minutes. In reality, I can usually paint a plant in a day; occasionally I can paint two. Some, like the Bull Banksia, took much longer.

My motivation to keep going

My enthusiasm for painting from nature has never diminished. The prospect of painting 400 plants, which grew to a final tally of 498 by the end of our project, was certainly a little daunting. However, I do believe that to identify and illustrate the plants of our area, at this moment in history, is important not only for our, but for future generations. New plants will surely be found, but sadly, many will be lost forever. I would love to see artists get together with botanists or plant enthusiasts in other areas, to produce more books like this one. I really believe that the more people are in touch with nature, and come to know and love our plants, the more chance we will have of preserving our unique heritage for our sons and daughters.

Gymnospermae

Zamiaceae

Macrozamia

G. *macros* = large + genus *Zamia*

1 *Macrozamia riedlei*, Zamia Palm

After Anselme Riedle (1775–1801), gardener at the Paris botanical gardens; member of the scientific staff on board Baudin's expedition on the Geographe that, together with Hamelin on the Naturaliste, explored the Western Australian coast, 1800–04

This plant belongs to a primitive family superficially similar to the palms, though not related. It now has only 80 species worldwide, but was once much larger and more widely distributed. It is dioecious, having separate male and female plants, each with large, attractive cones. The plants are generally about 1 m tall, with fronds up to 2 m long. The male cones develop and shed pollen in spring, while the seeds are shed from the female cone in autumn, but only germinate after further development the following year. The seeds are toxic, but were still an important food source for indigenous people who buried them for eight to nine months, rendering them safe to eat. Three species of Zamia grow on various soil types, mostly in Jarrah forest, but also woodland and heath. *Macrozamia riedlei* is found between Eneabba and Israelite Bay.

Podocarpaceae

This very ancient family of coniferous trees and shrubs is found mainly in the southern hemisphere and includes the Celery Top Pine and Tasmania's famous Huon Pine. It is related to the Taxaceae, the Yew family, of the northern hemisphere.

Podocarpus

G. *pous, podos* = foot + *carpos* = fruit; refers to the fleshy stalk of the fruit

2 *Podocarpus drouynianus*, Wild Plum, Emu Plum, Koolah

After Edourd Drouyn de Lhuys (1805–81), president of the French Imperial Acclimatization Society

Emu Plum is a strange shrub that grows as clumps of leafy stems that may reach up to 3 m arising from the nodes of underground creeping stems. The male cone appears in spring a year after fire and the female matures the following year on an enlarged receptacle that becomes an edible, but fairly tasteless fruit. Emu Plum grows mainly on sandy or lateritic soils in forest and woodland. It is found in the southern Jarrah and adjacent Karri forest regions.

A female *Macrozamia reidlei* behind a male plant, with *Pimelea rosea* growing between.

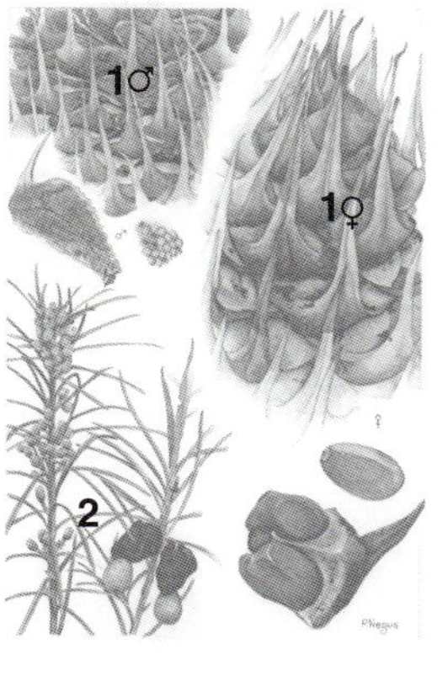

♂
♀
P.Negus

Angiospermae

The angiosperms are divided into two major groups: the **monocotyledons**, which have only one seed leaf (the first produced leaf); and the **dicotyledons**, which have two. Monocotyledons generally have leaves with parallel veins while the leaves of dicotyledons have netted veins. Another major difference is in the stems. In monocotyledons stems only reach a certain diameter, whereas the stems of dicotyledons generally continue to grow in thickness throughout life and many produce bark. The floral parts of monocotyledons are generally arranged in threes or multiples of three while those of dicotyledons vary, but are rarely in threes.

Many of the monocotyledonous families such as the grasses, rushes and sedges have in common a highly modified and reduced floral form. This is because they are adapted for wind pollination. They are usually a drab green or brownish colour since they do not need to attract pollinators. These features make them particularly difficult to identify and we make no apology for leaving the vast realm of the Poaceae, Restionaceae and Cyperaceae for another volume when we have learnt more about them! Here we have included just a very small selection of distinctive species.

Monocotyledons

Poaceae, Grasses

We have chosen to illustrate just three grasses, two native and one weed species, that are distinctive and well-known members of our coastal vegetation.

Spinifex

L. *spina* = spine, thorn + *-fex* = maker; refers to the pungent leaves of the originally named species

Do not become confused between this genus of primarily coastal and, in Australia, soft-leaved species, and the various desert 'Spinifex' species. The desert species are the spiny ones, but they belong to *Triodia,* a different genus altogether.

1 *Spinifex hirsutus*, Hairy Spinifex **05917484; 05917476**

L. *hirsutus* = shaggy, bristly

This is a tussock-forming perennial grass of coastal sand-dunes, with broad, silky-silvery leaves. Its thick rhizomes help to bind the shifting sand and it is a prime coloniser of the fore-dunes. It grows to less than 1 m high, but may spread for several metres along the sand. Its distribution is from the Perth region all around the southern coast of WA.

Flowers: Nov.–Jan.

2 *Spinifex longifolius*, Beach Spinifex **05626838; 05624819**

L. *longus* = long + *folium* = leaf

Beach Spinifex differs from the previous species in having longer, narrower leaves that are hairless and green. Its rounded, ball-like female flowers break off and are rolled along the beach by the wind. It is generally found further back from the beach than *S. hirsutus* and extends much further north from Albany up to the Kimberley coast.

Flowers: Aug.–Jan.

Ammophila

G. *ammos* = sand + *phileo* = to love

3 **Ammophila arenaria*, Marram Grass

L. *arenarius* = pertaining to or growing in sand

This is obviously a plant of the sand-dunes! Marram was introduced from Europe for use in dune reclamation projects, a role for which it has been very successful. While it is vigorous in colonising the dunes, it gives way to other plants once the conditions for plant growth have become more favourable. It occurs between Perth and Bremer Bay.

Flowers: Oct.–Jan.

Errata: On the key vignettes -

p. 14: Nos. 1 & 2 should be reversed (*Spinifex hirsutus* & *S. longifolius*)

p. 132: Nos. 3 & 4 should be reversed (*Hypocalymma angustifolium* & *H. robustum*)

Restionaceae, Southern Rushes

This is a family of sedge-like, perennial plants that are found in the southern hemisphere, mainly in South Africa and Australia. They are well-adapted to life in nutrient-poor and seasonally waterlogged habitats. The Restionaceae family reaches its greatest Australian diversity in the swamps of the far south-west and forms an important element of the vegetation, although its members generally go unnoticed by most of us. They have tough, wiry stems, along which are obvious nodes, each with a small, often papery leaf-scale. They are generally dioecious, having separate male and female flowers that may look markedly different. This, added to the superficial similarities of many species, makes them notoriously difficult to identify. They are, nevertheless, fascinating plants that, in many cases, are elegant and striking. Many have great future potential for rehabilitation projects on degraded lands, or as ornamentals, soil stabilisers and habitat for fauna around dams and waterways.

Only a very small sample of the species found in our area is included in this book.

Anarthria

G. *anarthros* = without joints

1 *Anarthria prolifera* **05626897; 05626900**

L. *proles* = progeny + *fero* = to bear; *bot.* progeny from offshoots

A rather sinuous, twining plant to about 60 cm with lovely red-brown flowers. The male and female flowers are easily distinguished, though similar. It forms large clumps on sandy soils in heath and woodlands. It is widely distributed around the coastal plain from Perth to Esperance.

Flowers: June–Oct.

2 *Anarthria scabra* **05624207; 05624215**

L. *scaber* = rough to the touch

This is one of the most prominent of the southern rushes, with broad, sedge-like leaves and striking, colourful flowers that emerge from reddish-brown sheaths. The styles of the female plant may be yellow or pink. *A. scabra* covers large expanses of peaty sand flats and winter-wet swamps all along the south coast. Margaret River is about its northernmost limit.

Flowers: Sept.–Nov.

Lyginia

L. *lyginos* = of willow or withy

3 *Lyginia imberbis* **05797489; 05797756**

L. *imberbis* = beardless

A tall rush with a tufted habit, growing up to 1 m tall, this species is widespread on dry, or seasonally wet sand plains. Here, it is common in the Nindup Plain and Calgardup areas. It is found from the Murchison River south, right around the coastal plain to Israelite Bay.

Flowers: July–Nov.

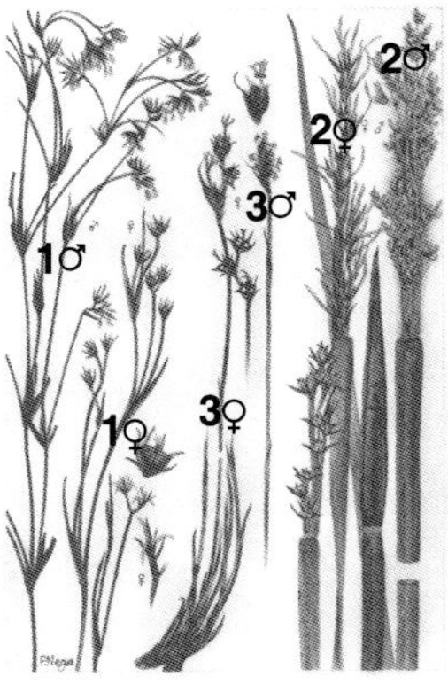

♂
♀
♀
♂
♀
♀
♂
P.Negus

Meeboldina

After Alfred Meebold (1863–1952), a German collector who travelled widely and left a collection of some 25,000 specimens to the Munich State Herbarium.

It can be very difficult to distinguish *Meeboldina* species, as the separate male and female plants often look radically different and many species grow together in the winter-wet swamps that are their favoured habitat.

1 *Meeboldina scariosa*, Velvet Rush **05623898; 05623944**

L. *scariosus* = thin, dry, membranous

This is a lovely tall rush growing to about 1 m, with handsome pink to red-brown flowers. The male has pendulous, nodding spikelets, while the female flower head is tufted. Two female flower heads are illustrated, the pinker one being younger. It is common in seasonally inundated swamps and heath on sand over peat soils from Gingin to Cape Riche.

Flowers: Nov.–Feb.

Loxocarya, Twine Rushes

G. *loxos* = oblique + *caryon* = nut

2 *Loxocarya cinerea*, Chinese Puzzle **05624223; 05623987**

L. *cinereus* = ash-coloured, grey

There are several species of *Loxocarya* in our area and all look very similar. This one is found growing in extensive clumps, often along roadsides, on sandy or lateritic soil. It is about 50 cm high and male and female plants appear similar from a distance, though their flowers are quite different. The female flower is minute, but its delicate lilac styles are beautiful to look at through a hand-lens. *L. cinerea* is a common species, extending from the Darling Range south of Perth to Augusta and along the south coast to Albany.

Flowers: Sept.–Oct.

Hypolaena

G. *hypo* = under + *chlaina* = cloak; refers to the glumes at the base of the spikes

3 *Hypolaena exsulca* **05795206**

L. *ex* = without + *sulcus* = furrow; the Eastern States species, *H. fastigiata,* has striate stems

This is a common little species. The nodding heads of the male plants are easily seen in sandy jarrah forest or heathland once you start to look for them. It is a dainty plant, up to 70 cm tall and is widespread throughout the south-west from Eneabba to east of Esperance.

Flowers: Sept.–Dec.

4 *Hypolaena pubescens* **05797624; 05797616**

L. *pubes* = soft, downy hair

This plant may belong to the same genus as the previous one, but it looks quite different. It is hairy, many-stemmed and tufted and may grow up to 1 m tall, though it is often low and straggly. The youngest emerging stems are particularly hairy. This species grows on peaty sand on dry or winter-wet flats and swampy depressions and is found along the coastal plain from Gingin to Albany.

Flowers: Oct.–Mar.

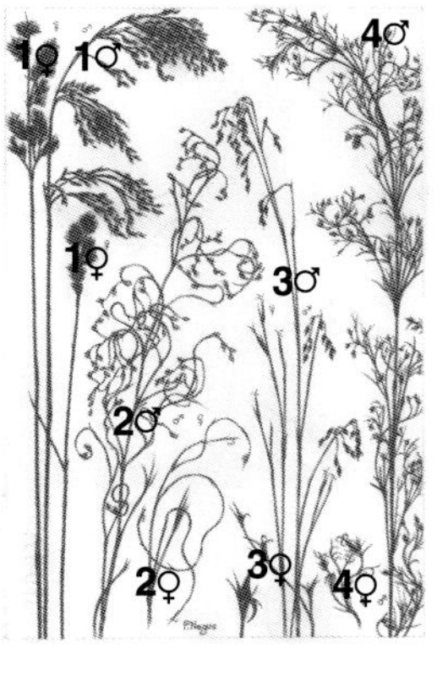

♀
♂
♀
♀
♂
♂
♀
♀
♂
P.Negus

Cyperaceae, Sedges

This is another very large, worldwide family, of which many species are found in wetland habitats. It is well-represented in the south-west corner and we have included a small sample of the more distinctive and easily recognised species. Sedges generally have brown flowers and flattened or circular leaves, with flowers and leaves all arising from the base of the plant. The fruits of many are bright, shining nuts.

Lepidosperma, Sword Sedges

G. *lepidos* = scale + *sperma* = seed

This genus of tufted, perennial sedges is poorly known and there may currently be more unnamed than named species! It is, however, fascinating to examine them closely and to realise that there are, in fact, so many different species that vary tremendously in size, leaf shape and in the characteristics of their inflorescence.

1 *Lepidosperma effusum*, Spreading Sword Sedge **05795044**

L. *effusus* = spread out, straggling

This is the very large, broadleaved species often seen along stream-banks and in moist depressions in the forests of the wetter south-west. It grows up to 2.5 m high and nearly as wide, on peaty sand and loam soils. The inflorescence is long and weeping.

Flowers: April–June

2 *Lepidosperma tetraquetrum*, Square Sedge **05795222**

L. *tetra* = four + *quetrus* = angled

Another large, but stiffer sedge, to 2 or 3 m, with more slender flowers than the previous species, and easily distinguished by its angular stems and leaves. It is often seen along the roadsides and is common in the forests of the south-west, between Perth and Albany.

Flowers: Nov.–Mar.

3 *Lepidosperma squamatum* **05611849**

L. *squama* = scale; *bot.* having small, scale-like leaves or bracts

This is a smaller species, growing to about 1 m. It grows on lateritic, sandy soil in Jarrah forest, where it is, as illustrated, a quite robust, broadleaved plant. It is also found on vegetated sand-dunes, where it forms extensive mats of long, much more slender-leaved plants. This species is widely distributed between Kalbarri and Israelite Bay.

Flowers: most of year, but especially Mar.–June

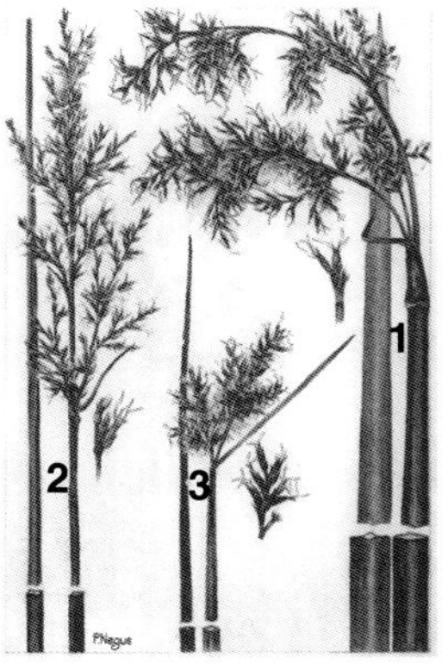

P.Negus

1 *Lepidosperma gladiatum*, Coastal Sword Sedge **05772443**

L. *gladius* = sword

A tough, broadleaved species, with chunky flower heads, common on coastal dunes, limestone and along coastal creek-lines. It is rather smaller than *L. effusum* and *L. tetraquetrum,* growing to about 1.5 m. It occurs around the south-west coast from Geraldton to east of Esperance.

Flowers: Nov.–Mar.

2 *Lepidosperma pubisquamatum* **05611822**

L. *pubes* = soft, downy hair + *squama* = scales

This is one of the many smaller *Lepidosperma* species that are all rather difficult to distinguish. It has smaller, narrower leaves than *L. squamatum* and grows only to about 50 cm. Common in Jarrah forest and woodland on lateritic or sandy clay soils from Eneabba to Bremer Bay.

Flowers: May–June

Tetraria

G. *tetras* = four; some species have styles branched into four.

3 *Tetraria octandra*, Eight-anthered Sedge **05624118**

G. *octo* = eight + *aner, andros* = man; *bot.* stamen

This attractive, leafy sedge looks rather like *Schoenus grandiflorus* (p. 26), but has eight distinctive, wispy yellow anthers that are initially twisted in pairs. It can be over 1 m tall and grows equally well in swamps or on rocky hillsides. It is found between Eneabba and Albany.

Flowers: July–Oct.

Cyathochaeta

G. *cyathos* = cup + *chaite* = long hair; refers to the long bristles

4 *Cyathochaeta avenacea* **05795052**

From the genus, *Avena* (wild oats) + *-aceous*; the dried flowerspikes somewhat resemble those of wild oats.

This is a slender, tufted sedge up to 1.5 m tall, that generally goes unnoticed amongst the other bushland grasses and sedges. It is most apparent in summer, after the very tall, grass-like flower spike has dried. Although scattered, it is a common plant of the damper Jarrah forest and woodland and along swamp edges. It occurs between Jurien Bay and Hopetoun.

Flowers: Nov.–Jan.

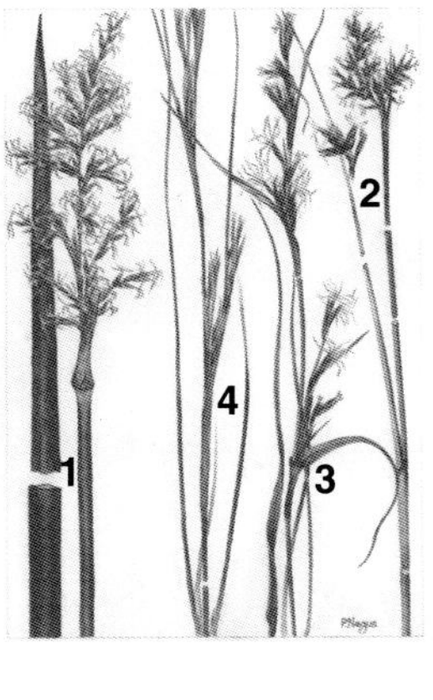

P.Negus

Gahnia

After Henricus Gahn (1747–1816), Swedish botanist and friend of Linnaeus. He wrote about grasses.

1 ***Gahnia decomposita*****,** Saw Sedge **05626609**

L. *de-* (neg. prefix) + *compositus* = brought together, united; ie. separated, divided

A very large, tussocky perennial, growing up to 3 m tall, in swamps, streams and on seasonally-wet flats. It has masses of long, drooping sharp-edged leaves and tall, slender flower heads, that also droop over as they develop. It likes to grow in the damp forests of the lower south-west, between Perth and Albany.

Flowers: Sept.–Nov.

2 ***Gahnia trifida*****,** Coast Saw Sedge **05624835**

L. *trifidus* = three-forked

A similarly robust, tussocky sedge, a little smaller, to about 1.5 m, but with globular clusters of flowers on long spikes. It grows on sand in saline habitats, such as the Deadwater Lakes at the mouth of the Blackwood River and is widely distributed along the coast and inland between Kalbarri and Cape Arid.

Flowers: Aug.–Oct.

Evandra

G. *eu* = well, abundance of + *aner, andros* = man (*bot.* stamen); this plant has 12 or more stamens

3 ***Evandra aristata*** **05797519**

L. *arista* = the awn or beard of an ear of corn; *bot.* awned, bearded

This is a lovely sedge of sand plains and swamplands. It grows in tussocks with delicate, nodding flower heads on slender stems that reach to over 2 m. It is mainly found between Busselton and Albany.

Flowers: Oct.–Dec.

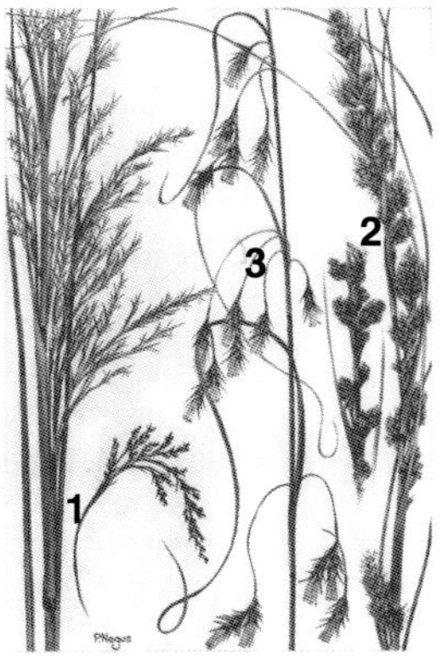

P.Negus

Schoenus

G. *schoinos* = rush

1 *Schoenus grandiflorus*, Large-flowered Bog Rush **05627001**

L. *grandis* = large + *flos, floris* = flower

Unusually, amongst sedges, the Bog Rushes have leafy stems. This one is a robust and handsome tufted perennial, with erect flower stems that usually grow to about 50 cm, though they can reach 1 m or more. It is found on dunes and sand-plains around the coast from Kalbarri to Cape Arid.

Flowers: May–Aug.

Mesomelaena

G. *mesos* = middle + *melaino* = to turn black; refers to the dark or black involucral bracts

2 *Mesomelaena tetragona*, Semaphore Sedge **05611881**

G. *tetragonus* = four equal angles, square; the flower stems are square towards the top

This is probably our most recognisable sedge, with its two long bracts extending beyond the flower as if signalling in semaphore. The dark brown or black flower heads look dead for most of the year, but come alive when the yellow stamens burst forth from new flowers each autumn. It grows in clumps up to 1 m tall in Jarrah woodland and sandy heath, usually in moister locations. Widespread from Perth to east of Esperance.

Flowers: April–June

Isolepis

G. *isos* = equal + *lepis* = scale

3 *Isolepis nodosa*, Knotted Club Rush **05772478**

L. *nodosus* = full of knots; *bot.* bearing knot-like swellings or nodes

This sedge should be familiar to beach-goers, as it is a common inhabitant of the dunes and the edges of creeks and rivers near the sea. It is salt-tolerant and hardy, making it a valuable dune stabiliser. It grows to about 1 m high and is found from Shark Bay all around the south-west coast to Esperance.

Flowers: Nov.–Mar.

Baumea, Twig Rushes

4 *Baumea vaginalis*, Sheath Twig Rush **05795095**

L. *vagina* = sheath; *bot.* sheathed

Twig Rushes are not the most prominent of plants and you may be forgiven for not registering their existence. However, like many of the Cyperaceae and Restionaceae, they are actually very important components of the vegetation in wetland habitats and there are several species growing in our area. This one is a robust, tufted perennial, growing to 1.5 m and is commonly found in winter-wet depressions and along watercourses, between Gingin and Albany. If you are lucky enough to catch it in flower, you will see that it is in fact a rather lovely plant.

Flowers: Sept.–Nov.

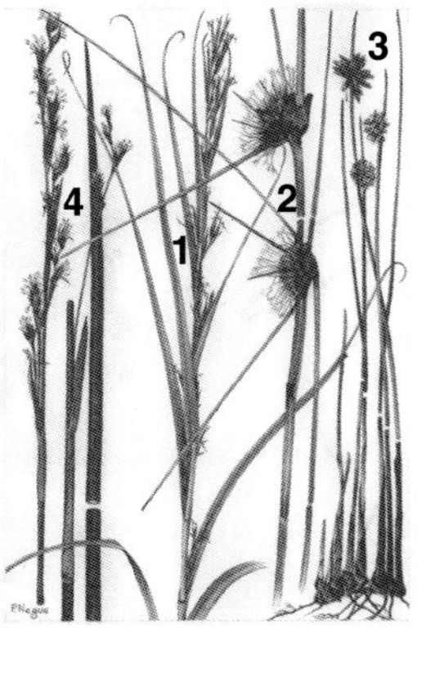

P.Negus

Juncaceae, Rushes

Members of the rush family usually have a tussocky growth habit and often have round, hollow leaves. They are found in damp locations. The three illustrated here have been positively identified, but there are sure to be other species growing in the area.

Juncus

L. *juncus* = rush

1 ***Juncus kraussii*** subsp. ***australiensis*****,** Sea Rush **05799171**

A tall perennial to a little over 1 m that grows in dense stands in brackish swamps, estuaries and on saline flats all along the coast from Shark Bay to Esperance. It is common along the edges of streams such as Boodjidup and Calgardup Brook where they open to the sea, and along the tidal stretch of Hardy Inlet. The flower is a deep red-brown.

Flowers: Oct.–Jan.

2 ****Juncus microcephalus*** **05797535**

G. *micros* = small + *cephale* = head

This is an introduced species, native to South Africa, and found between Perth and Albany. It grows to about 2 m and is found in all types of moist locations such as roadside drains, springs and streams. A native species, *J. holoschoenus,* looks very similar, but has yellow-brown rather than red-brown fruit capsules.

Flowers: Nov.–Feb.

3 ***Juncus pallidus*****,** Pale Rush **05797810**

L. *pallidus* = pale

This is a robust rush, growing to 2 m, found on various soil types in swamps and along watercourses. A very common species in the area and throughout the whole south-west.

Flowers: Nov.–Feb.

Typhaceae, Bulrushes

Typha

G. *typhe* = reed-mace, *T. angustata*

There is only one genus with two species in WA. The more delicate *T. domingensis,* is native to northern Australia, but may have spread into the south-west, while the taller, more robust *T. orientalis,* is native to Eastern Australia.

4 ****Typha orientalis*****,** Bulrush **05799155**

L. *orientalis* = pertaining to the east

A robust perennial, growing up to 4.5 m tall, it is found in streams and wet depressions and is a vigorous coloniser of disturbed wetlands around the south-west coastal plain.

Flowers: Nov.–Jan.

6 ***Calectasia grandiflora*** subsp. ***'southern' (H. Demarz*** **546*)***
Blue Tinsel Lily, Star of Bethlehem (described page 30)

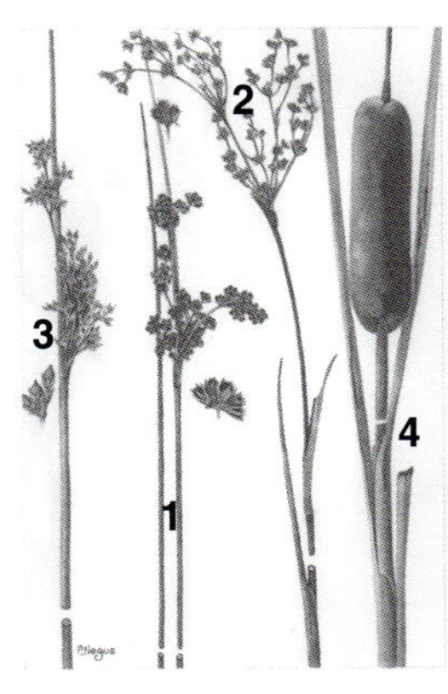

P.Negus

Xanthorrhoeaceae, Grasstrees

This family now comprises only the genus *Xanthorrhoea*, of which there are currently nine species recognised in WA. They vary in the presence or absence, and length, of a trunk, and the relative lengths of their flower spike and scape, the stem-like flower stalk.

Xanthorrhoea

G. *xanthos* = yellow + *rheo* = to flow; refers to the resin that flows from the stem.

The trunks of *Xanthorrhoea* species are interesting, as they comprise layer upon layer of old leaf bases, glued together with resin and arranged around a central stem.

1 *Xanthorrhoea brunonis* subsp. ***semibarbata*** **05772206**

After Robert Brown (1773–1858), naturalist on Matthew Flinders' scientific expedition to Australia in the Investigator, 1801–03. He stayed in Australia until 1805 and was undoubtedly the most distinguished early collector of WA flora. He also described and named many WA plants collected by others.

This is one of the grasstrees without a trunk, but its long, stout flowering stalk is up to 1.5 m tall. It is common on sandy heath and laterite and is found mostly between Perth and Augusta. The illustration shows the flower spike detail.

Flowers: Oct.–Dec.

2 *Xanthorrhoea preissii*, Grasstree, Balga

After Johann August Ludwig Preiss (1811–83), who collected over 2,700 species in WA, 1839–41.

This is probably the best known of the grasstrees, with a trunk, often branched and generally blackened by fire, up to 3 m high. The flower spike is much longer than its scape. It is common in forest and woodlands, on lateritic or sandy soils, from Geraldton to Albany. It flowers most profusely after bush-fires. The whole plant is illustrated.

Flowers: any month, but usually late spring.

3 *Xanthorrhoea gracilis*, Graceful Grasstree **05772486**

L. *gracilis* = thin, slender

Its tuft of leaves is sparser, but this grasstree may not be recognised until it sends up its flower stalk, which is much more slender than that of *X. brunonis* or *X. preissii*. The spike is much smaller and looks quite different. It is common on loam or gravel soils in Jarrah forest between Perth and Albany.

Flowers: Oct.–Jan.

Dasypogonaceae

The plants in this family were previously placed in Xanthorrhoeaceae. They comprise a diverse group and vary widely in form and habit. Ongoing studies in this family are likely to result in further reclassification.

Kingia

After Phillip Parker King (1791–1856), hydrographer and naturalist and first Australian to reach the rank of rear-admiral. He sailed to the north-west coast of WA with Cunningham and Roe. His father, Philip Gidley King (1758–1808), was governor of NSW at the time of the Investigator voyage of Flinders and Brown and was of great assistance to them.

4 *Kingia australis* **05612004**

L. *australis* = southern

This genus is endemic to south-west WA and has only one species. It is a tall grasstree that can grow to 8 m and individuals are known to have reached ages of between 750 and 900 years. Like *Xanthorrhoea* species, it has a fibrous trunk surrounded by layers of tightly-packed leaf-bases, but it has drumstick-shaped flower heads. Its leaves, unlike those of *Xanthorrhoea* species, are finely hairy and glisten in sunlight. It grows on sandy or clay loam soils in woodland or open heath. Widespread between Perth and Albany.

Flowers: July–Aug.

Dasypogon

G. *dasys* = thick with hair, rough + *pogon* = beard

There are three species, endemic to the south-west. Again there are similarities with the grasstrees, but *Dasypogon* species have drumstick flower heads and their leaves are flat or inrolled.

5 *Dasypogon hookeri*, Pineapple Bush **05611997**

After Sir William Jackson Hooker (1785–1865), director of Kew Gardens.

A magnificent tree-like perennial, growing to 3 m, with a crown of broad pineapple-like leaves from which emerge several flowering stems. It is often seen growing in groups on sandy, gravel or clay soils in Jarrah forest. It has a fairly localised distribution between Donnybrook and Augusta.

Flowers: Aug.–Dec.

Calectasia

G. *calos* = beautiful + *ectasia* = stretching out; refers to the star-shaped perianth-lobes

6 *Calectasia grandiflora* subsp. ***'southern' (H. Demarz* 546*)*,** **05625009**
Blue Tinsel Lily, Star of Bethlehem (illustrated page 28)

L. *grandis* = large + *flos, floris* = flower

I have so far found this superb plant in only one location in our area, just north of Margaret River on thin gravel soil over granite. I hope it occurs here more frequently, as it is a truly exquisite species! Its electric blue perianth segments turn a deep maroon as they age. It is an open, woody sub-shrub, growing only to about 50 cm. The species is found between Perth and Esperance, with the 'southern' variant mainly found in the area between Walpole, Mt Barker and Albany.

Flowers: Aug.–Oct.

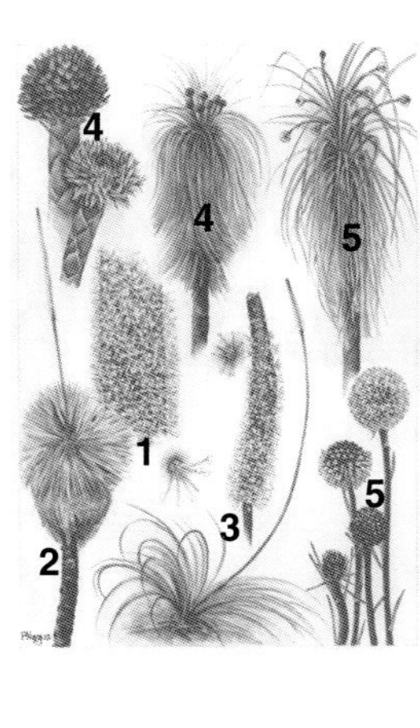

P.Negus

1 *Dasypogon bromeliifolius*, Drumsticks **05772192**

From the genus *Bromelia* + L. *folius* = leaf

This plant is quite different from Pineapple-bush, being a robust, but ground-hugging, tufted perennial with grass-like leaves, generally less than 1 m high. The flowers are borne on long hairy stems. It is often found in large clumps in sandy heath or woodland. Common between Jurien and Bremer Bay.

Flowers: Oct.–Dec.

Lomandra, Mat Rushes

G. *loma* = border of a robe + *aner, andros* = man (*bot.* anther); the anthers of some species have borders

Members of this genus are dioecious, having separate male and female plants and are mostly small, tussock-forming perennials.

2 *Lomandra nigricans* **05611962**

L. *niger* = black, *bot.* becoming black, drying black

This little herb has a tuft of grass-like leaves and a flower spike generally not more than 50 cm high. The male flowers are white with purple markings and are on short stalks, while female flowers are stemless and often greenish in colour. This is one of several species that are difficult to distinguish from each other. It grows on sandy or lateritic soils in Jarrah forest or woodland and is widespread from Perth to Esperance.

Flowers: June–Sept.

3 *Lomandra purpurea*, Purple Mat Rush **05771811**

L. *purpureus* = purple

This is a taller species, with erect, slender flower spikes to over 1 m. The yellow anthers against the purple background of the petals make this a striking flower. It is common in Jarrah forest on sandy and lateritic soils. It is found from Perth to Albany.

Flowers: Oct.–Nov.

4 *Lomandra caespitosa*, Tufted Mat Rush (illustrated left) **05624940**

L. *caespes* = mound or thatch of turf; *bot.* tufted

A small, but interesting species, with a tiny spike of cream or yellow flowers only about 10 cm long. The illustration shows only the male flower. The whole plant is only about 30 cm tall. Under ground level, though, this plant is tough and woody. Although difficult to find, it is again quite common in Jarrah forest and woodland on sandy or lateritic soils between Moora and Albany.

Flowers: July–Oct.

Acanthocarpus

G. *acantha* = thorn, prickle + *carpos* = fruit; refers to the spines on the valves of the fruit

5 *Acanthocarpos preissii*, Prickle Lily **05611954**

After Johann August Ludwig Preiss (1811–83) – see *Xanthorrhoea preissii.*

This is an interesting little plant that is easily overlooked, but is actually very prevalent on the limestone sands around the coast. It is a straggly, tufted perennial with drab grey-green foliage and spiny leaves, but the clusters of flowers are sweetly scented and the seeds inside their spiny fruit capsule are sweet to eat when ripe. This plant grows vigorously in disturbed sites. It is found along the coastline from Carnarvon south to Augusta.

Flowers: April–June

4 *Lomandra caespitosa*, Tufted Mat Rush

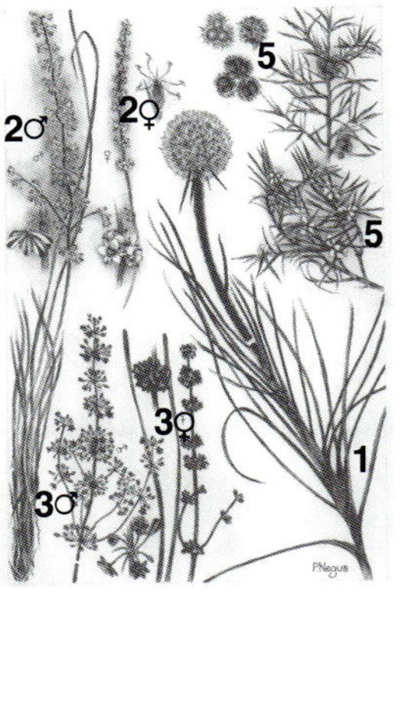

♂
♀
♀
♂
P.Negus

Anthericaceae, Lilies

Plants that used to be grouped together in the very large Lily family (Liliaceae), have now been split into several smaller groups. The Anthericaceae is the largest of these in WA and the whole group is likely to undergo further reclassification in the future.

Caesia

After Federigo Cesi (1585–1630), Italian naturalist, said to have been the first to discover the spores of ferns.

1 *Caesia micrantha,* Pale Grass Lily **05625025**

G. *micros* = small + *anthos* = flower

This is a slender lily, growing up to about 50 cm, with small white flowers that are easily overlooked. It grows in a variety of habitats, here mainly in Jarrah forest or Peppermint woodland. It is widespread between Geraldton and Ravensthorpe.

Flowers: Oct.–Nov.

Hodgsoniola

After John Hodgson (1799–1860), Victorian politician, mayor of Melbourne and patron of the arts.

2 *Hodgsoniola junciformis* **05626722**

L. *juncus* = a rush + *forma* = shape, appearance

A delicate perennial with attractive mauve flowers that peep out amongst the sedges and rushes in the winter-wet swamps and sand plains. It grows to about 50 cm on peaty sand. After flowering, the petals twist into a spiral. This little lily has a limited distribution between Bunbury, Manjimup and Augusta.

Flowers: Oct.–Dec.

Johnsonia

After Thomas Johnson (*c* 1597–1644), English apothecary and naturalist.

3 *Johnsonia lupulina,* Hooded Lily **05625165**

L. *lupulus* = hop-plant + *-inus* = like; refers to the resemblance of the flower heads to those of the European hop

This is a tufted perennial, whose multiple, nodding flower heads are very showy and conspicuous in the bushland. It grows to about 80 cm and likes sand or gravelly soils in open forest or woodland. It survives well on disturbed sites such as roadsides. Look closely at the flower heads to see the tiny individual flowers, protected by the papery pink and white bracts. Found between Busselton and Albany.

Flowers: Oct.–Dec.

Sowerbaea

After James Sowerby (1757–1822), English botanical artist.

4 *Sowerbaea laxiflora,* Purple Tassels, Vanilla Lily **05624029**

L. *laxus* = loose, open + *flos, floris* = flower

The beautiful little individual star flowers of Purple Tassels tend to open just two or three at a time, giving only a hint of what might be! The plants of this slender, tuberous perennial are scattered through bushland of the coastal plain between Dongara and Augusta. It is found on sandy soil in woodland or open heath, usually amongst a colourful suite of herbs.

Flowers: Sept.–Nov.

Tricoryne

G. *treis* = three + *coryne* = club; the fruit is divided into three club-shaped nutlets

5 *Tricoryne elatior,* Yellow Autumn Lily **05797748**

L. *elatior* = taller

This is a trailing, perennial herb with several flowers on long slender pedicels. It grows through the vegetation and, by supporting itself on other plants, can grow to 1 m or so. It is common on sandy or lateritic soils in woodland or heath and is widely distributed throughout the state.

Flowers: Dec.–Mar.

6 *Tricoryne humilis* **05625017**

L. *humilis* = low, small

This small herb, only about 20 cm high and easy to overlook, is one of eleven species of *Tricoryne* currently recognised. It is found on sand or gravel, often in damp places, but is not very common here. Found between Geraldton and Albany.

Flowers: Sept.–Nov.

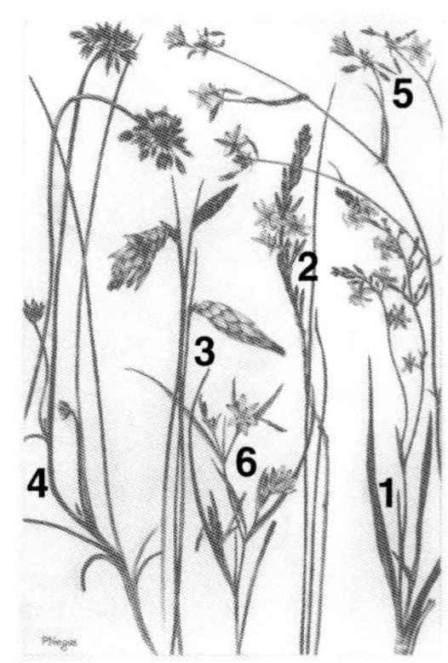

P.Negus

Thysanotus, Fringed Lilies

G. *thysanotos* = fringed; refers to the fringed perianth segments (petals and sepals)

While their delicate purple flowers may appear similar, plants of this genus takes on a multitude of forms and there are a surprisingly large number of species, 49 currently recognised in WA. Many have tuberous roots that were much sought after for food by indigenous people. A delightful feature of this group is the way the flowers close at night with the fringe neatly tucked away within the calyx.

1 *Thysanotus arenarius* **05515459**

G. *arenarius* = pertaining to, or growing in sand

This is a rather limp, trailing perennial with long leafless stems, reaching up to about 70 cm, usually when supported by other vegetation. As the name suggests, it grows in sand over limestone. It is found all along the west coast from Dongara to Augusta.

Flowers: Oct.–Dec.

2 *Thysanotus dichotomus*, Branching Fringed Lily **05772028**

G. *dicha* = in two ways + *tomios* = cut up; *bot.* forked, branching into two

This is an interesting species, somewhat similar to the last, having long, leafless branches, but its growth habit seems somehow more decisive! It seems more robust and its branches become tangled, often forming a loose mound, from which the flowers emerge. It has basal, linear leaves that wither before the flowers appear. Widespread throughout the south-west from Shark Bay to Walpole.

Flowers: Nov.–Jan.

3 *Thysanotus manglesianus*, Twining Fringed Lily **05771986**

After James Mangles, captain RN (1786–1867), naturalist and patron of botanical collecting.

A twining, leafless perennial that can reach 2 m in height. The flowers, however, are quite small. It grows on a variety of soils in woodland and heath and is widespread throughout the state.

Flowers: Aug.–Nov.

4 *Thysanotus multiflorus*, Many-flowered Fringed Lily **05515505**

L. *multus* = many + *flos, floris* = flower

This species must be one of our most handsome lilies. The terminal heads of bright purple flowers, only two or three opening at a time, contrast dramatically with the silver-grey buds and stems. It is a stout plant with long, linear, basal leaves. Grows on sandy, laterite or granitic soils throughout the forested south-west.

Flowers: Oct.–Jan.

5 *Thysanotus tenellus* **05625076**

L. *tenellus* = somewhat tender or dainty

This is a tiny, delicate plant, reaching only about 20 cm high. It has a tuft of wispy, grass-like leaves. You need to search hard for this one, which grows on sandy soils, often on granite outcrops or in swampy areas. It is found between Jurien Bay and Albany.

Flowers: Sept.–Oct.

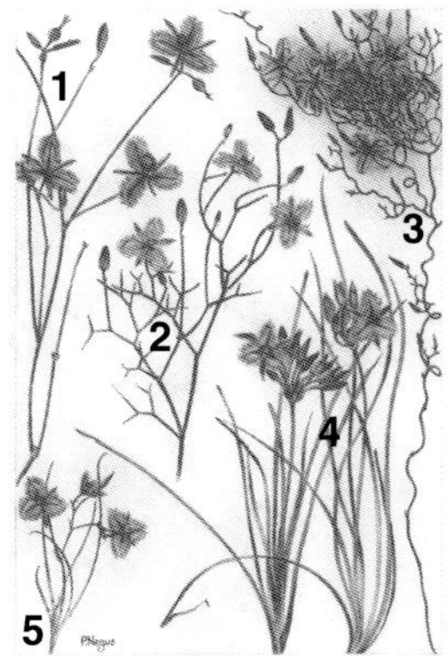

P.Negus

Agrostocrinum

G. *agrostis* = grass + *crinon* = lily

1 ***Agrostocrinum stypandroides,*** Blue Grass Lily, False Blind Grass **05749913**

From the genus *Stypandra* + *-oides* = like (see below)

A large, tufted herb with striking blue flowers that tend to open in the morning sun. It grows to about 1 m on clay, laterite, or peaty sand, in forest or swampy heath. Scattered through the south-west mainly between Perth and Esperance.

Flowers: Oct.–Dec.

Chamaescilla

G. *chamai* = on the ground + *scilla* = sea-onion or squill

2 ***Chamaescilla corymbosa*** var. ***corymbosa,*** Blue Squill **05749921**

L. *corymbus* = a cluster of flowers + *-osus* = abounding in; *bot.* corymbe: flower head with a flat top, the lower flower stalks being longer than those above

A small perennial herb with an erect stem to about 20 cm and a basal rosette of narrow, linear leaves. It is common in Jarrah forest, Sheoak woodland and heath, from Kalbarri to Esperance.

Flowers: Aug.–Dec.

Phormiaceae

Dianella

After Diana, Roman goddess of hunting, created queen of the woods by Jupiter + dim. *-ell-*; the original species was found in woods in southern France and called *dianelle des bois.*

3 ***Dianella revoluta,*** Blueberry Lily, Flax Lily **05749891**

L. *revolutus* = rolled back or down (the leaves)

Though this lovely species is a sturdy plant, its flowers are small and dainty. It can grow to 1.5 m in sheltered situations, but is common here on coastal sand over granite or limestone in windswept heathland where it is pruned much lower. Its fruit is a deep blue berry. It is widespread through the inland and south-west.

Flowers: Sept.–Jan.

Stypandra

G. *stype* = tow: flax fibres made ready for spinning, tousle (of hair) + *aner, andros* = anther; anthers are woolly-tomentose

4 ***Stypandra glauca,*** Blind Grass **05749905**

L. *glaucus* = bluish-grey or -green (the leaves)

A very handsome plant up to 1 m tall that grows in colonies on and around granite outcrops. It is apparently toxic to stock and can cause blindness in sheep; hence the common name. Blind Grass is widespread through the south-west.

Flowers: Aug.–Nov.

Colchicaceae

Burchardia

After Johann Heinrich Burckhard (1676–1738), German botanist.

This is another group of useful plants that have nutritious, tuberous roots.

5 ***Burchardia multiflora,*** Dwarf Burchardia **05624045**

L. *multus* = many + *flos, floris* = flower

This is a small but stout little plant, standing erect about 30 cm high with broad, stem-clasping leaves. The flowers have a lovely pinkish tinge. It grows on peaty sand or clay in swamps and amongst the moss on granite outcrops. Found between Eneabba and Albany.

Flowers: Aug.–Oct.

6 ***Burchardia umbellata,*** Milkmaids **05474523**

L. *umbella* = parasol, umbrella; flower head umbrella-shaped, oldest flowers to the outside

Taller and more slender than the last species, this common herb is found scattered through open forest and woodlands on a variety of soils. Milkmaids have flower stalks up to 60 cm high. It is widespread from Geraldton to Albany.

Flowers: Aug.–Oct.

Wurmbea

After F. von Wurmb, merchant and secretary of the Academy of Sciences of Batavia in the 18th century.

Whereas the flowers of *Burchardia* are in umbels, those of *Wurmbea* are solitary or in a spike.

7 ***Wurmbea dioica*** subsp. ***alba,*** Early Nancy **05749948**

G. *dis* = double + *-oicos* = house; *bot.* dioecious, having male and female organs on separate plants; L. *alba* = white

This is a small, delicate plant usually less than 30 cm high, with two or more flowers on the erect stem. It grows on sandy soils nestled amongst heathland shrubs or on mossy granite outcrops and is found from Kalbarri to Bremer Bay.

Flowers: July–Sept.

Boryaceae

Borya, Pincushions

After Jean Baptiste George Marie Bory de Saint-Vincent (1780–1846), French naturalist attached to Baudin's expedition to Australia.

These are a group of small, hardy, tufted perennials, mainly found on granite outcrops, where they have to withstand extremes of waterlogging in winter and desiccation in summer.

8 ***Borya scirpoidea*** **05625033**

From the genus *Scirpus,* a group of tough grasses, + *-oides* = like

This is a spreading, tufted little plant, only 10 or 20 cm high, found on damp, sandy flats and granite. It is found between Dongara and Albany.

Flowers: Sept.–Oct.

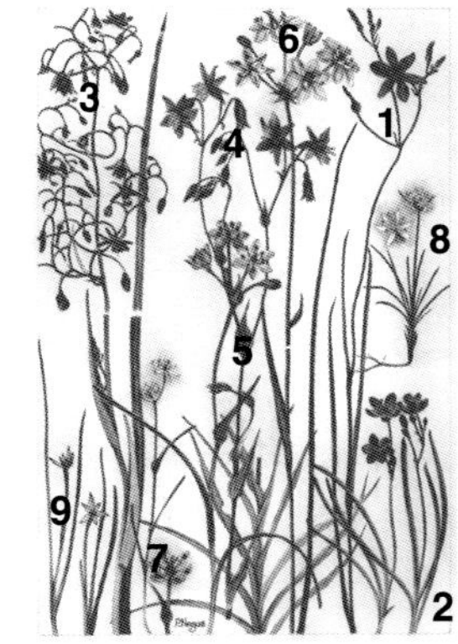

P.Negus

Iridaceae, Irises

The introduced species in this family, some of which are illustrated later in the weeds section, unfortunately far outnumber the natives. There are several beautiful native irises, however, in just two genera. They are distinguished from the lilies by having an inferior rather than a superior ovary.

Patersonia

After William Paterson (1755–1810), one of the earliest botanists of the colony; a friend of Robert Brown, to whom, and to Sir Joseph Banks, he sent specimens.

Patersonia flowers are very delicate and wither quickly. They emerge one at a time from a 'purse' of buds, formed by enclosing brown bracts.

1 ***Patersonia occidentalis***, Purple Flags **05749964**

L. *occidentalis* = western

A lovely, tufted perennial, with broad, spreading leaves and flowers up to 70 cm high. It grows on sand, clay or laterite and favours a variety of habitats such as winter-wet areas, dunes or granite outcrops. It is widespread between Kalbarri and Cape Arid.

Flowers: Sept.–Jan.

2 ***Patersonia umbrosa*** var. ***xanthina***, Yellow Flags **05474434**

L. *umbrosus* = very shady, growing in shade; G. *xanthos* = yellow

A tall perennial, to nearly 1 m, with narrow leaves, this species is most obvious when flowering. It frequents sandy or gravelly soils in Jarrah forest where large numbers of plants often make very showy displays. The flowers generally only come out on sunny days. This species is most common in the lower south-west, between Perth and Albany.

Flowers: Aug.–Nov.

Orthrosanthus, Morning Irises

G. *orthros* = dawn, early morning + *anthos* = flower

3 ***Orthrosanthus polystachyus***, Many Spike Orthrosanthus **05772044**

G. *polys* = many + *stachys* = ear of corn, flower spike

This is a magnificent plant deserving of a place in any garden. Its flower spikes can be up to 1 m tall, with a profusion of sky-blue flowers. It grows on sand or laterite in forest and woodland, mainly in the southern half of the area. Its general distribution is from Busselton to Albany.

Flowers: Oct.–Dec.

Xyridaceae

Xyris

Taken from the name for *Iris foetidissima*

4 ***Xyris roycei*** **05797500**

After Robert Dunlop Royce (*b* 1914–), curator of Perth Herbarium, 1960–74

This is one of eighteen species of yellow, iris-like plants found mainly in the wetlands of the south-west. Individually, they are not particularly spectacular plants, but are very showy when flowering en-masse in the swamps. This one has quite large flowers and can grow to nearly 1 m. It is currently known only from a small area between Margaret River and Walpole.

Flowers: Oct.–Nov.

Philydraceae

Philydrella

Genus: *Philydrum* + dim. *-ell-* (*Philydrum* = G. *phileo* = to love + *hydor* = water)

5 ***Philydrella pygmaea***, Butterfly Flower **05625149**

G. *pygmaios* = dwarf

This tiny little plant, just a few centimetres tall, is common but difficult to find unless in full flower. It inhabits damp, swampy areas and also granite outcrops where, in both cases, it is found in large numbers. The single stems have several buds, but each flower usually appears singly. It is a unique plant with two petals and only one stamen. Common between Geraldton and Esperance.

Flowers: Sept.–Nov.

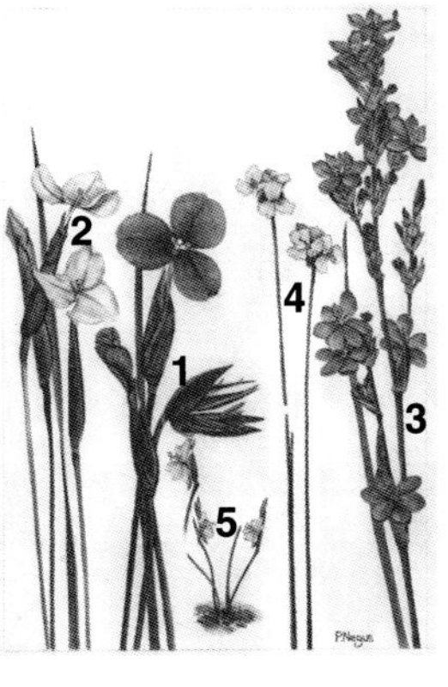

P.Negus

Hypoxidaceae

Hypoxis

G. *hypo* = under + *oxys* = sharp, pointed; may refer to the capsule that is elongated and contracted at the base

9 *Hypoxis gardneri*, Yellow Star (illustrated page 39) **05627087**

After Charles Austin Gardner (1896–1970), Government Botanist; he discovered eight new genera and some 200 new species!

A tiny lily-like plant with 6 petal-like flower parts, which is abundant in wet places, such as the moss swards on granite outcrops. It has a fairly restricted distribution, being recorded mainly from the Darling Range between Perth and Collie. Its occurrence in the Margaret River-Augusta area makes a considerable southerly extension to its range.

Flowers: July–Aug.

1 *Hypoxis occidentalis* var. ***quadriloba*,** Yellow Star (illustrated below) **05626994**

L. *occidentalis* = western

This little lily looks almost identical to the last, except for its 4 instead of 6 petal-like parts. It is found in the same kind of habitat, and indeed, the two species are often seen growing together. It has a wider distribution throughout the wetter areas of the south-west from Perth to Bremer Bay.

Flowers: July–Oct.

Haemodoraceae

This family does have relatives elsewhere and may well have more South African than Australian members. However, it is at its most diverse in south-west Australia. The kangaroo paws are almost certainly the best known of this group of stout perennial herbs, but there are also several other interesting genera as well.

Anigozanthos, Kangaroo Paws

Jacques Julien Houton de le Billardiere, who named this genus, left no explanation as to his derivation for the name. Some suggested alternatives have been:

G. *anisos* = unequal + *anthos* = flower – the flower's unequal corolla

G. *anicho* = elevated + *anthos* = flower – flowers held on elevated stems

G. *anoigo* = to open, undo + *anthos* = flower – open branching of the flower stems

or, from a "fancied resemblance to the plumes of horses and mourning coaches"

Meanings are not always evident and easily become lost in the mists of time!

Kangaroo paws and cottonheads (see next page) are covered in a thick mat of woolly hairs that helps them to conserve moisture.

2 *Anigozanthos manglesii*, Mangles Kangaroo Paw, Red and Green Kangaroo Paw **05772060**

After Robert Mangles (*d* 1860), brother of James, horticulturalist in whose garden the type specimen was grown

Best known and loved of the kangaroo paws, our WA state emblem grows well in the area and can be seen in quantity along Caves Road, Cowaramup Bay Road and elsewhere when in flower. The flower stems are up to 1 m tall and it grows in sandy soils in heath or woodland between Shark Bay and Mt. Barker.

Flowers: Sept.–Nov.

3 *Anigozanthos flavidus*, Tall Kangaroo Paw, Evergreen Kangaroo Paw **05797772**

L. *flavidus* = yellowish

This is the most prominent kangaroo paw in our area and we actually have a 'Margaret River' colour variety that is much redder than normal. The tall plants, with flowers sometimes over 2 m tall, are evident in large clumps along wet roadside ditches, riverbanks and swamps throughout the area. The flowering stems look superb when highlighted against the early morning or late afternoon sun. It is found between the South-west Capes and Two Peoples' Bay.

Flowers: Nov.–Jan.

4 *Anigozanthos viridus*, Green Kangaroo Paw

L. *viridus* = green

We have cheated slightly by including this handsome species, as I have so far only found it growing north of the Shire boundary. However, as it also occurs in the Scott River region east of Augusta, there is every chance that it does or has grown in suitable habitats here as well. The flower stems are up to 1 m tall and it grows in open heath in sandy or clay, winter-wet depressions and occurs along the coastal plain as far north as Nambung National Park.

Flowers: Aug.–Oct.

1 *Hypoxis occidentalis* var. ***quadriloba*,** Yellow Star

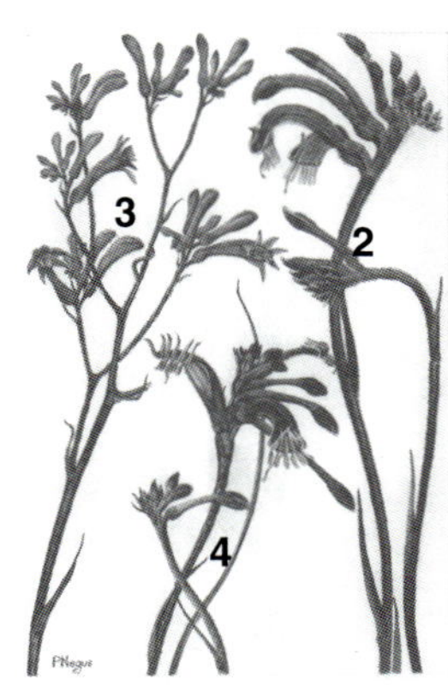

P.Negus

Conostylis, Cottonheads

G. *conos* = cone + *stylos* = column, *bot.* style

There are about forty-five currently recognised species in the south-west, many of these with varieties or subspecies, so this is a large and variable genus. Most are yellow, with similarly-shaped flowers, so you need to examine them closely in order to distinguish them.

1 ***Conostylis aculeata*** subsp. ***aculeata***, Prickly Conostylis **05461014**

L. *aculeatus* = furnished with prickles!

A very variable species with ten subspecies. The diagnostic features are the ascending spines along the leaf edges that are generally easy to see. It is a grass-like perennial with flattened leaves, low and spreading, to about 50 cm and is found on a variety of soils over limestone or laterite, in heath, woodland or forest. It occurs between Jurien Bay and Albany.

Flowers: Aug.–Nov.

2 ***Conostylis laxiflora*** **05625300**

L. *laxus* = wide, loose, open + *flos, floris* = flower

The pale yellow flowers of this species open out more than many other *Conostylis* and are quite large. Their clusters nestle in the midst of the leaves of this low plant that only grows to about 40 cm. It is found in Jarrah forest on sandy clay in moist locations, mostly in the Capes area. A very similar and more widespread species, *C. serrulata,* has longer peduncles (flower stalks) and grows on drier, gravelly soils.

Flowers: Sept.–Oct.

3 ***Conostylis setigera*** subsp. ***setigera***, Bristly Cottonhead **05749956**

L. *seta* = bristle, stiff hair + *gero* = to bear

This is another small, tufted perennial, only about 30 cm high with distinctive white hairs that grows on sand or laterite soils in dry heath or woodland. It is common between Moora and Bremer Bay.

Flowers: Sept.–Oct.

Phlebocarya

G. *phleps, phlebos* = vein + *caryon* = nut

4 ***Phlebocarya ciliata*** **05771900**

L. *cilium* = eye-lash; *bot.* fringed with fine hairs

A tufted perennial with long, sedge-like leaves and loose panicles of flowers. It grows in large clumps on sandy soil, in winter-wet depressions or drier Banksia woodland, occasionally in Jarrah forest, around the coastal plain from Jurien Bay to Albany.

Flowers: Sept.–Nov.

Tribonanthes

G. *tribon* = cloak + *anthos* = flower; refers to filament appendages, that rise above the anthers

Though quite unrelated these furry white flowers are reminiscent of Alpine edelweiss!

5 ***Tribonanthes australis*** **05623995**

L. *australis* = southern

A stout, handsome plant, to about 50 cm, with flowers covered in dense, felty hairs. It grows in damp situations, often amongst moss swards on granite outcrops. It is found from Geraldton down the west coast and eastwards to Albany.

Flowers: Aug.–Sept.

P.Negus

Haemodorum, Blood Roots

The name used by Theoprastus and Pliny for *Orobanche cruenta,* a parasitic plant. Its derivation here is not substantiated, but is likely from: G. *haima* = blood + *dorum* = gift

These plants are often overlooked in the bush, despite their size, maybe because they blend in with the ever-present, fire-blackened twigs and dead grasses. Their bright red leaf bases are a striking feature of this genus and their rootstock is edible. The three species here are widely distributed between Geraldton and Albany.

1 *Haemodorum laxum* **05797594**

L. *laxus* = wide, loose, open

This is a tall-stemmed plant to well over 1 m, with thick, basal green leaves and brown-black widely-branching stems. It is found on sandy clay soils, usually in seasonally damp locations in Jarrah forest or open woodland.

Flowers: Nov.–Dec.

2 *Haemodorum simplex* **05771927**

L. *simplex* = simple, undivided

This is a much smaller plant than the other two, growing to only about 60 cm. The flower is fragrantly scented and it is has rather beautiful orange, bulging three-lobed fruits. Grows on sand or clay in winter-wet, swampy sites and around granite outcrops such as those just north of Augusta.

Flowers: Sept.–Oct.

3 *Haemodorum spicatum* **05797780**

L. *spica* = point, flower spike

The tall, straight black stems of this species, up to 1.4 m high, are evident once your eye is tuned to seeing them. They are found in heath or forest and often grow in groups along the road verges.

Flowers: Nov.–Dec.

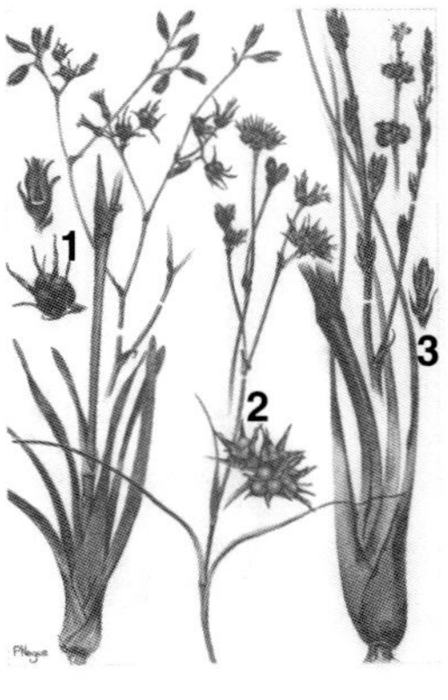

P.Negus

Orchidaceae, Orchids

The orchids are a fascinating group of plants. Their flower structure is unique, having the stamens and style fused to form a structure called the column, and at least one petal, known as the labellum, modified to create a landing stage for the insects that pollinate them. Most are specialised for pollination by only one specific type of insect and their adaptations to this end are, in many cases, quite bizarre!

This is the world's largest plant family, with over 30,000 species. The terrestrial orchids of the south-west are perhaps not as large and glamorous as their tropical counterparts, but their forms are just as intriguing. More than 300 species have been identified in the south-west region and at least 100 are probably present in the area covered by this book. The names of many of them change as their relationships become better studied and their precise identification requires specialist knowledge. We have included just a sample here and have grouped them by flowering season.

Orchids of late spring and summer

Thelymitra, Sun Orchids

G. *thelys* = female + *mitra* = cap; refers to the hooded column

1 ***Thelymitra fuscolutea,*** Chestnut Sun Orchid

L. *fuscus* = dark + *luteus* = yellow

Very similar to the Leopard Orchid, *T. benthamiana* that is also found here, this one has smaller individual flowers, more prominent markings and a later flowering time. It is about 30 cm tall and grows in clumps around the margins of winter-wet swamps, but it is also found in drier Jarrah forest, between Perth and Albany.

Flowers: Dec.–Jan.

2 ***Thelymitra*** aff. ***macrophylla,*** Scented Sun Orchid

G. *macros* = large, long + *phyllon* = leaf

This is one of several closely related blue sun orchids that have yet to be fully described. It has a tall, stout spike to about 60 cm, often with a large number of flowers that are spectacular when they are all open at once. They do need sunshine, however, and patience is needed to catch them at their best. Grows in Jarrah forest, on sand plains and around swamp edges.

Flowers: Oct.–Nov.

Elythranthera, Enamel Orchids

G. *elytron* + cover + *anthera* = *bot.* anther: the column wings hide the anthers

3 ***Elythranthera brunonis,*** Purple Enamel Orchid

After Robert Brown (1773–1858), naturalist on Matthew Flinders' voyage and renowned early collector of WA flora

A beautiful and distinctive orchid that is common in Jarrah forest and heathland throughout the south-west from Kalbarri to Israelite Bay. It is about 30 cm high, with bright, shining petals.

Flowers: Sept.–Nov.

E. emarginata, Pink Enamel Orchid, is also common here but is more scattered and flowers a little later.

Prasophyllum, Leek Orchids

G. *prason* = leek + *phyllon* = leaf

4 ***Prasophyllum brownii,*** Christmas Leek Orchid

After Robert Brown (1773–1858), naturalist on Matthew Flinders' expedition and the most distinguished early collector of WA flora

Although this resembles some of the other tall leek orchids, this one is easily recognised by its brown-striped petals and sepals and its late flowering period. It can grow to 1.2 m and is found in Jarrah and Karri forest of the lower south-west between Perth and Albany.

Flowers: Dec.–Jan.

Cryptostylis

G. *cryptos* = hidden + *stylos* = column; the column is hidden by the base of the labellum

5 ***Cryptostylis ovata,*** Slipper Orchid

L. *ovum* = egg; *bot.* ovate (leaf)

This orchid is quite famous for its pollination method of sexually attracting the male Ichneumon wasp. The insect tries to mate with the flower and, in doing so, picks up pollen that is then transferred to the next flower it visits. Many orchids are now known to have a similar relationship with their pollinators. The prominent leaves of this orchid can be found growing in colonies all year round, but only a few plants produce flowers. The flower stems are up to 70 cm tall and have several flowers that open one at a time. It grows under thick coastal scrub, woodland or in Jarrah or Karri forest all through the lower south-west.

Flowers: Dec.–Mar.

Praecoxanthus

6 ***Praecoxanthus aphyllus,*** Leafless Orchid

G. *a-* = without + *phyllon* = leaf

The flower of this slender species is about 40 cm high and the flower looks somewhat like a spider orchid, a group in which it was formerly placed. However, unlike these, it flowers in late summer and only non-flowering plants bear a leaf that is shiny, small and oval. This orchid is found in sandy soil in Banksia or Sheoak woodlands and low heathland along the south coast.

Flowers: Mar.–Apr.

P.Negus

Autumn/Winter Orchids

Leporella

L. *lepus, leporis* = hare + *-ell-* dim.

1 *Leporella fimbriata*, Hare Orchid

L. *fimbriatus* = fringed

The leaves of this orchid are seen much more easily than the flowers. They grow in large colonies on sandy soils in Banksia, Jarrah or Sheoak woodlands, or underneath coastal scrub. It flowers most readily after fire, though one or two flowers usually appear every year. The stems are only about 25 cm high, but the flowers are very persistent, lasting a long time after the fertilised ovary begins to swell, so it is worth looking for them well into winter. This orchid has a wide distribution from north of Kalbarri right along the south coast to well beyond Israelite Bay.

Flowers: Mar.–June

Eriochilus, Bunny Orchids

G. *erion* = wool + *cheilos* = lip; the labellum is pubescent (finely hairy)

2 *Eriochilus dilatatus*, White Bunny Orchid

L. *dilatatus* = enlarged, widened (the labellum)

This orchid is very variable, with several recognised sub-species. It is quick to flower after the first autumn rains and its bright flower spike can be spotted in Karri forest or tucked under the tall shrubs of the coastal heath.

Flowers: April–June

Prasophyllum (see previous page)

3 *Prasophyllum parvifolium*, Autumn Leek Orchid

L. *parvus* = small + *folium* = leaf

There are two forms of this orchid: the early flowering one (April–June), shown here, is light green and white and strongly scented, while the later flowerer (June–Aug.) has green-purple flowers and is odourless. The early Scented Autumn Leek Orchid is more common, but both are found here, scattered through the coastal heathland, sheltering beneath the shrubs. Both are dainty little plants up to about 30 cm tall.

Pterostylis

G. *pteron* = wing + *stylos* = column

This group of orchids ensures cross-pollination by enticing the pollinating insect, a small gnat or mosquito, into the heart of the flower where it becomes trapped. The only exit is via a tunnel, formed by the winged column and the labellum, that guides the insect first past the stigma, onto which any previously gathered pollen is deposited and then past the anthers from which new pollen is collected.

4 *Pterostylis rogersii*, Curled-tongue Shell Orchid

After Richard Sanders Rogers (1862–1942), a leading authority on Australian orchids. He described 79 new species.

This lovely little species is surprisingly common along the limestone ridge, growing in large colonies in Peppermint woodland and heath. Its flowers are quite easy to spot once you become aware of them. Its colour ranges from pink to greenish- or reddish-brown. It is found around the coast from Bunbury to Israelite Bay.

Flowers: May–July

5 *Pterostylis* aff. *nana*, Snail Orchid

L. *nana* = dwarf

The Snail is one of our very common winter orchids, growing up to 30 cm and found equally at home amongst the leaf litter of the Jarrah or Karri forest, or in the shelter of shrubs of the coastal heath. There are currently several distinctly recognised species and varieties of snail orchids growing in the area, but their differences are often difficult to pick in the field and they are yet to be formally named.

Flowers: June–Sept.

6 *Pterostylis vittata*, Banded Greenhood

L. *vitta* = a band worn on the head; *bot.* vittate, having longitudinal stripes

A robust species that is common in forest or heath. It can grow to around 50 cm. Found mainly between Perth and Albany, but spreading around the coast to Dongara and Esperance.

Flowers: June–Sept.

Cyanicula

G. *cyaneos* = blue

7 *Cyanicula deformis*, Blue Fairy Orchid, Blue Beard

L. *deformis* = irregularly shaped

It is surprising to find this colourful little orchid flowering in the middle of winter, in some most inhospitable spots near the coast. It generally favours damp situations, but is commonly seen here amongst dense low heath along the Leeuwin–Naturaliste Ridge. It is usually only about 15 cm high. Common throughout the south-west from north of Kalbarri to Israelite Bay.

Flowers: July–Sept.

Corybas, Helmet Orchids

Corybas was the first published name for this genus and is therefore upheld, although it has no relevance to this orchid. It was actually a pirated version of the true, intended name: Corysanthes, G. *corys* = helmet + *anthos* = flower.

8 *Corybas recurvus*, Forest Helmet Orchid

L. *recurvus* = bent backwards (the flower)

The leaves of this orchid can often be seen in large colonies on the forest floor amongst deep leaf litter and moss, but only a few plants usually flower. However, if you look hard, you will find them once your eye is accustomed to their colour and shape. On the Leeuwin–Naturaliste Ridge, they are often found at the entrances to caves, where the atmosphere is always humid. An enlarged dorsal sepal forms the prominent hood of this fascinating little plant, protecting its large labellum, which hides the column. The other sepals and petals are much reduced. When fertilised, the flower stalks elongate up to 30 cm, a characteristic that may aid seed dispersal. Found between Bunbury and Albany.

Flowers: July–Sept.

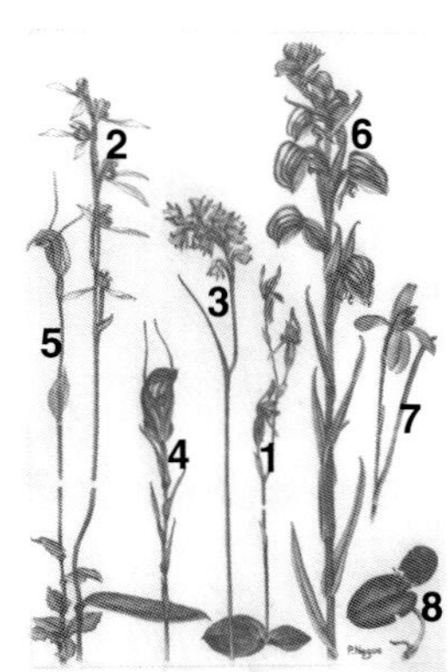

P.Negus

Orchids of Early Spring

Diuris, Donkey Orchids

G. *dis* = double + *oura* = tail; refers to the two hanging, lateral sepals

1 *Diuris* aff. *amplissima*, Giant Donkey Orchid

L. *amplus* = large, splendid

One of several donkey orchids that occurs in the area, this magnificent orchid is found on the sandy limestone soils of the Leeuwin–Naturaliste Ridge, where it shelters under the Peppermint and other tall shrubs of the coastal heath. It grows very tall, sometimes nearly to 1 m.

Flowers: Sept.–Oct.

Pterostylis (see previous page)

2 *Pterostylis recurva*, Jug Orchid

L. *recurvus* = curved backwards (the lateral sepals)

A handsome orchid growing to around 60 cm, often pushing up through the surrounding undergrowth. Each stem may have up to four flowers. It is usually found in Jarrah forest and occurs between Geraldton and Israelite Bay.

Flowers: Aug.–Sept.

3 *Pterostylis barbata*, Bird Orchid

L. *barbatus* = bearded

This attractive orchid is one of several similar-looking, and closely-related, species that grow in the south-west. It grows in Jarrah forest and Sheoak woodland from Bindoon to Albany and is about 30 cm high.

Flowers: Aug.–Oct.

A little different, but still confusing to the lay person, are the *Pterostylis turfosa* group of bird orchids that have a much more densely-hairy labellum. One of these, the Late Bird Orchid, is quite common here and is often found in Peppermint woodland flowering, as the name suggests, in later spring (mid-Oct.–Dec.)

Lyperanthus

G. *lyperos* = sad + *anthos* = flower

4 *Lyperanthus serratus*, Rattle Beaks

L. *serratus* = saw-edged, toothed; refers to the prominent white calli on the labellum

The popular name for this orchid comes from the rattle produced by the flowers when they are shaken. There are up to ten flowers, twisting around each stalk. This is a large and common orchid, to about 50 cm high, but one that tends to grow amongst thick vegetation where it is often overlooked. Widespread from north of Perth to Israelite Bay.

Flowers: Sept.–Oct.

Cyrtostylis

G. *cyrtos* = curved + *stylos* = column; refers to the incurved column

5 *Cyrtostylis huegelii*, Midge Orchid, Mosquito Orchid

After Carl Alexander Anselm, Baron von Hügel (1796–1870), Austrian traveller and naturalist

Found in colonies amongst the moist litter of the forest floor, or under dense heathland scrub, the leaves of this orchid are easily seen, but the flowers are more cryptic and you generally need to look hard to find them. Pollinating insects such as small flies or gnats are attracted to a trail of nectar that spills from the base of the narrow labellum down a central groove. Occurs from Kalbarri to east of Esperance.

Flowers: July–Sept.

Drakaea, Hammer Orchids

After Miss Drake (most active 1818–47), English botanical illustrator

These orchids are pollinated by small male Thinnid wasps that are attracted by a scent emitted by the orchid, similar to that of a female wasp. The labellum also looks a little like the female, at least to the male wasp. The male wasp tries to fly off with the labellum and in doing so brushes against the column and picks up a package of pollen that it deposits on the next flower it visits.

6 *Drakaea glyptodon*, King-in-his-carriage

glyptodon = an extinct South American animal that resembled an armadillo. Perhaps refers to the glandular lobe of the labellum, or the general shape of the flower. The common name is much easier to visualise!

This slender orchid grows to about 30 cm and is found on sandy flats and around winter-wet swamps, in woodland or heath. It favours clearings and you can often find it along the side of sandy bush tracks. Found between Jurien Bay and Esperance.

Flowers: Aug.–Oct.

Paracaleana, Duck Orchids

G. *para* = beside + genus, *Caleana:* after George Caley (1770–1829), superintendent of the botanic gardens at St. Vincent

7 *Paracaleana nigrita*, Flying Duck Orchid

L. *nigritia* = black colour; refers to the dark glands along the labellum

This little one is difficult to find though it is actually quite common. Like the King-in-his-carriage, it prefers sandy plains and open woodland or heath. It has a tiny reddish leaf that may often be found before the flowers are discovered. It occurs from Bindoon, to east of Esperance.

Flowers: Sept.–Oct.

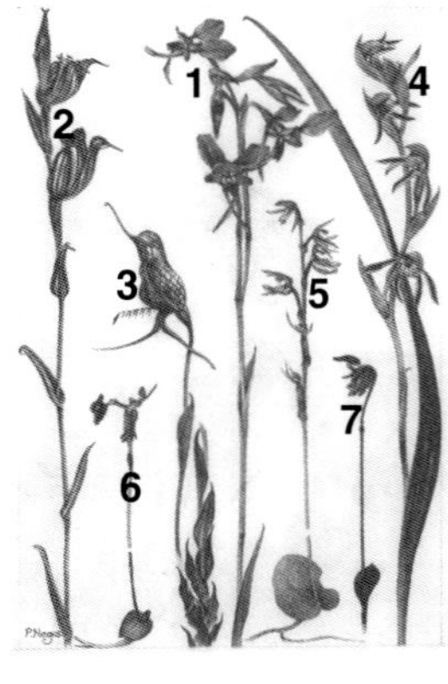

P.Negus

Caladenias – Spring-Flowering Spiders

As so many of our spring-flowering orchids are Caladenias, or spider orchids, we devote most of this page to them.

Caladenia, Spider Orchids

G. *calos* = beautiful + *aden* = gland; some species have colourful glands at the base of their column

1 *Caladenia applanata*, Broad-lipped Spider Orchid

L. *applanatus* = flattened

This is a robust, variably-coloured species, often having a deep red labellum and greenish petals It is commonly found on the coastal limestone, in low heath and Peppermint woodland. The flowers are usually found singly or only in twos or threes and it grows to about 50 cm high. It occurs in a thin coastal strip from Dunsborough to Albany.

Flowers: Sept.–Oct.

2 *Caladenia cairnsiana*, Zebra Orchid

After Adam Cairns (1802–81), Presbyterian minister in Melbourne, who encouraged various philanthropic and educational activities

Not all *Caladenia* species are spiders! This is a diminutive member of the genus, usually not more than 20 cm high and difficult to locate, but very charming. Look for it in sandy country in Jarrah forest or heath and on the thin soils around granite outcrops. It has quite a large range between Lancelin and Esperance.

Flowers: Aug.–Oct.

3 *Caladenia citrina*, Margaret River Spider Orchid

L. *citrinus* = lemon-coloured

Yes, we do have our very own spider orchid! It has so far only been found between Dunsborough and Forest Grove, where it occurs mostly on laterite or granite soils in Jarrah-Marri forest and most commonly flowers after fire. It looks superficially similar to the White Spider Orchid (*C. longicauda*) but its colour, its labellum fringe, and its stiffly-held petals and sepals are quite different.

Flowers: Sept.–Oct.

4 *Caladenia ferruginea*, Rusty Spider Orchid

L. *ferrugineus* = rust-coloured

A slender species with rusty-red flowers that have prominent glands at the ends of the petals and sepals. Several spider orchids have such glands and it is thought that they emit sexual attractants to lure pollinating insects. The Rusty Spider is found in Jarrah forest, on sandy or lateritic soils, between Perth and Mt. Barker, but is most abundant in our area.

Flowers: Sept.–Oct.

5 *Caladenia flava*, Cowslip Orchid

L. *flavus* = yellow

This is one of our most common and best-loved orchids. It grows in clonal groups that each has its own characteristics of size, shape and colour. It is found in a wide variety of habitats throughout the south-west between Geraldton and Israelite Bay.

Flowers: Aug.–Nov.

A subspecies, *C. flava* subsp. *sylvestris,* with slightly taller, pale yellow and white flowers and a long, slender leaf, is also found in forested areas and flowers a few weeks later. Both hybridise readily with other *Caladenia* species, such as Pink Fairies, Pink Fan, and the White Spider Orchid, producing some beautiful variations in flower shape and colour.

6 *Caladenia macrostylis*, Leaping Spider Orchid

G. *macros* = large + *stylos* = column

Its dense black calli along the labellum, knob-like clubs and leaping appearance make this a very distinctive spider orchid. It is only 20–25 cm high and is generally only found singly. It is found throughout the Jarrah and Karri forest regions of the south-west from north of Perth to Albany.

Flowers: Aug.–Nov.

Cyanicula, Blue Orchids

The blue orchids used to be grouped with *Caladenia,* but have recently been given their own genus.

7 *Cyanicula sericea*, Silky Blue Orchid

L. *sericeus* = silken

This lovely, tall blue orchid, up to 45 cm high, is most prominent after fire when singly, or in small groups, it stands out against the blackened ground and tree-trunks of the forest. Its main habitat is the Jarrah forest, where it prefers laterite and granite soils, but it is also found in Banksia woodland and granite outcrops, between Perth and Esperance.

Flowers: Aug.–Oct.

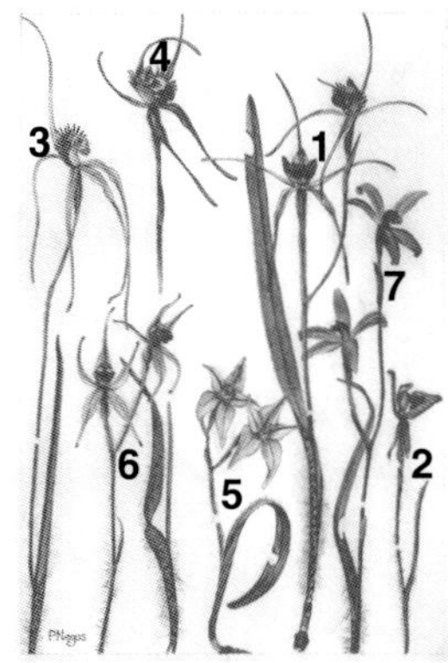

P.Negus

Dicotyledons

Casuarinaceae Sheoaks

The family name and that of the genus *Casuarina*, comes from the Malay word *casuari*, for the large emu-like bird, the cassowary, found in the jungles of SE Asia. The drooping branches of many species are seen to resemble the feathers of the bird. *Casuarina* species and those of the allied genus *Allocasuarina* (G. *allo* = other), generally have separate male and female flowers that are relatively insignificant, having no petals, though the mass of red styles on the female flowers are quite striking. Their leaves are reduced to tiny scales. However, their beautiful cones and the graceful forms of some of the larger tree species compensate well for these deficiencies.

In this area we have just two species of *Allocasuarina*.

1 *Allocasuarina fraseriana*, Sheoak **05795265**

After Charles Fraser (?1788–1831), first Superintendent of Sydney Botanical Gardens and NSW Colonial Botanist. He came to the Swan River region with James Stirling in 1827 to make a biological survey and his good report was used to initiate colonisation of the Swan River in 1829.

A. fraseriana is a small tree that can grow to 15 m in sheltered locations. It tends to be dominant, forming thickets, or sharing space with Jarrah or Banksia. Not much grows beneath its branches that drop millions of smothering needles onto the ground below. It occurs through the Jarrah forest to Augusta and along the south coast to Albany.

Flowers: much of the year.

2 *Allocasuarina humilis*, Dwarf Sheoak **05474698**

L. *humilis* = low

This species is a low, spreading shrub and is found on sandy soils in coastal and inland heath. It is widespread throughout the south-west. It can be dioecious (separate male and female plants) or monoecious (male and female flowers on the same plant).

Flowers: winter, spring, summer.

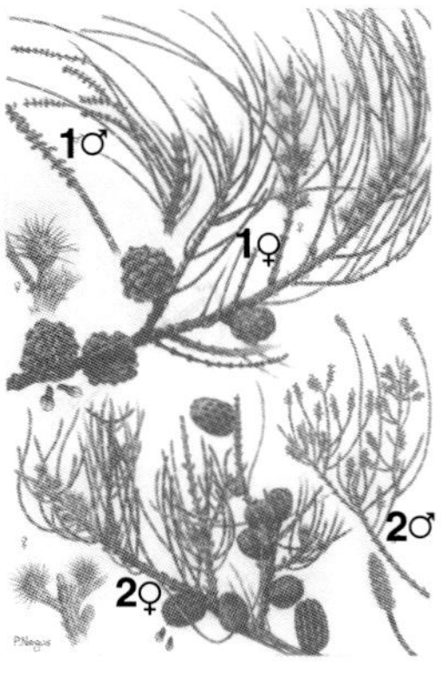

♂
♀
♀
♀
♂
P.Negus

Proteaceae

The Proteaceae is one of the largest and most important families represented in Western Australia. The family takes its name from the Greek god, Proteus, who was able to change his shape at will. There is a huge variety of flower shapes in this family, as you can see from the species that we have in our area. In some genera, such as *Adenanthos, Synaphea*, the flowers are separate, while some, such as *Banksia* and *Dryandra*, have flower heads containing a huge number of individuals that are fascinating to look at with a hand lens. The flowers may be regular or irregular, but all have a single whorl of four perianth segments.

Most members of the Proteaceae have a specialised system of roots called proteoid roots. These are tangled, fibrous networks of tiny roots that spread out near the soil surface, to make the most of scarce nutrients in a poor environment. It is this root system that makes members of the family very sensitive to dieback disease (*Phytophthora cinnamomi)* and to phosphate-rich fertilisers. They cannot cope with the nutrient overload!

Banksia

After Sir Joseph Banks (1741–1820), patron of natural sciences. He sailed around the world with Captain James Cook and collected the first known members of the genus at Botany Bay in 1770.

1 *Banksia grandis*, Bull Banksia (illustrated page 5)

L. *grandis* = large

Shrub or tree to 10 m. One of our prominent understorey trees, found on lateritic or sandy soils in Jarrah-Marri forest and coastal woodland. Distributed throughout the forested parts of the south-west. Its magnificent flower heads turn from greenish-yellow to golden as they age, and are hard to miss. Near Augusta, you will find it naturally wind-pruned to a low shrub on the coastal heath around Flinders Bay.

Flowers: Oct.–Jan.

2 *Banksia attenuata*, Candle Banksia, Slender Banksia

L. *attenuata* = narrowing to a point; refers to the leaves that narrow towards the base

Shrub or tree to 10 m, found mainly on the sandplains. There are many beautiful specimens along Cowaramup Bay Road towards Gracetown. It is widespread throughout the south-west from Murchison River to Ravensthorpe.

Flowers: Nov.–Jan.

3 *Banksia littoralis*, Swamp Banksia

L. *littoralis* = belonging to the sea-shore

Shrub or tree to 12 m, growing on black, peaty sand, in seasonally damp areas and along watercourses. Mainly found on the coastal plain between Jurien Bay and Albany. It grows with *B. attenuata* near Gracetown and looks similar, but can be distinguished by its less toothed leaves and its autumn flowering period.

Flowers: Mar.–May

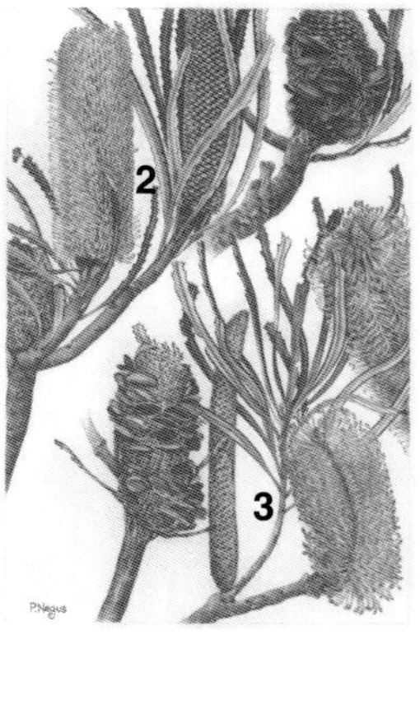

P. Negus

1 *Banksia ilicifolia*, Holly-leaved Banksia **05611970**

From *Ilex* (holly) + L. *folium* = leaf

Usually a small, spreading tree up to 5 m, with prickly leaves and flower heads more closely resembling those of *Dryandra*. It is, indeed, often confused with *Dryandra sessilis*, Parrot Bush, but though the latter does have hard fruiting follicles they are not embedded in a woody cone as they are in *Banksia* species. You will find *B. ilicifolia* on the sandplains behind the Leeuwin–Naturaliste Ridge on deep yellow sands, whereas *D. sessilis* favours the grey sands of the limestone ridge. The flowers of *B. ilicifolia* turn a beautiful pink as they get older. Unlike in most *Banksia* species, the cones of *B. ilicifolia* are very small, with only two or three almond-shaped seed capsules forming from any one flower head. It has a coastal distribution between Jurien Bay and Albany.

Flowers: any month, though mostly autumn.

Dryandra

After Jonas Dryander (1748–1810), librarian and fellow of the Linnean Society. Curator of Sir Joseph Banks' collections, therefore much concerned with Australian plants.

The genus *Dryandra* is closely related to *Banksia*, but bracts surround the flower heads. The fruiting follicles are typically leathery rather than woody and are not embedded in a woody cone, as in *Banksia* species.

2 *Dryandra sessilis* var. ***cordata***, Parrot Bush △ **05749859**

L. *sessilis* = sessile, without a stalk; refers to the leaves

A dense, prickly shrub or small tree to 3 m found on the coastal limestone. When in flower Parrot Bush is tremendously attractive to birds and the cacophony of squabbling honeyeaters and wattlebirds can be quite deafening! Parrot Bush is a common species, found all around the south-west coast, but this particular variety is only found between Cape Naturaliste and Walpole. It differs from *D. sessilis* var. *sessilis* in having much larger flowers and leaves that are also darker.

Flowers: July–Oct.

3 *Dryandra lindleyana* subsp. ***lindleyana***, Couch Honeypot **05624169**

After John Lindley (1799–1865), professor of botany, University College, London

This is a prostrate, ground-hugging shrub with underground stems and is usually no more than 30 cm high. It is found on sandy and lateritic soils in coastal heath and woodlands, between Geraldton and Albany. Another subspecies, *D. lindleyana* subsp. *sylvestris,* with broader leaves and individual leaf lobes also occurs here. This is the one found more usually in the forest.

Flowers: July–Sept.

P.Negus

Grevillea

After Charles Francis Greville (1749–1809), one of the founders of the Royal Horticultural Society.

Grevillea is a large and spectacular genus of shrubs and trees, many of which are now familiar garden plants. Over half the Australian species occur in WA, though we have only a few in this area.

1 *Grevillea bronwenae* **05456932**

After Bronwen Keighery, WA botanist and teacher

This gorgeous species is found mainly to the east of our area, but it does also occur around Cowaramup. Its distribution is restricted to the small area bounded by Busselton, Nannup and Margaret River. It is a small shrubto about 1 m with brilliant orange-red flowers, and it is found in moist locations in the Jarrah forest, on sandy soil over clay.

Flowers: July–Sept.

2 *Grevillea quercifolia*, Oak-leaf Grevillea **05474396**

From *Quercus* (oak) + *folium* = leaf

One of the most attractive small shrubs of the Jarrah forest, the unusual magenta flowers of this species make bright splashes of colour in early spring. It is a low, spreading shrub, less than 1 m high that prefers sandy, lateritic soils. It is found between Perth and Albany.

Flowers: Sept.–Oct.

3 *Grevillea pulchella* subsp. *ascendens*, Beautiful Grevillea **05749875**

L. *pulchella* = beautiful

A delicate, spreading shrub, less than 1 m high, with small creamy white flowers. This one can be difficult to find, hidden amongst taller vegetation. It occurs on sand or gravelly loam in Jarrah forest, mostly in the wetter south-west between Bunbury and Albany.

Flowers: Sept.–Nov.

4 *Grevillea trifida* **05460840**

L. *trifida* = (leaves) forked into three

Medium, spreading, spiny shrub to about 1.5 m, with masses of creamy white flowers all along the stems. Found in Jarrah forest, on sandy or gravelly soils generally on drier slopes, but sometimes also in swampy locations, between Pinjarra and Albany. In this area it is mostly found east of Bussell Highway.

Flowers: Aug.–Oct.

5 *Grevillea manglesioides* subsp. *manglesioides* **05611946**

After Captain James Mangles (1786–1867) naturalist + *-oides* = similar to

This tall, open shrub, about 3 m tall, with small, but attractive green and red flowers appears to have a fairly limited distribution between Busselton and Augusta and east to Manjimup. It grows along the old railway formation in gravelly soil above the Margaret River and a few other locations, but is more frequent in the Jarrah forests to the east.

Flowers: June–Nov.

5 *Grevillea manglesioides* subsp. ***manglesioides***

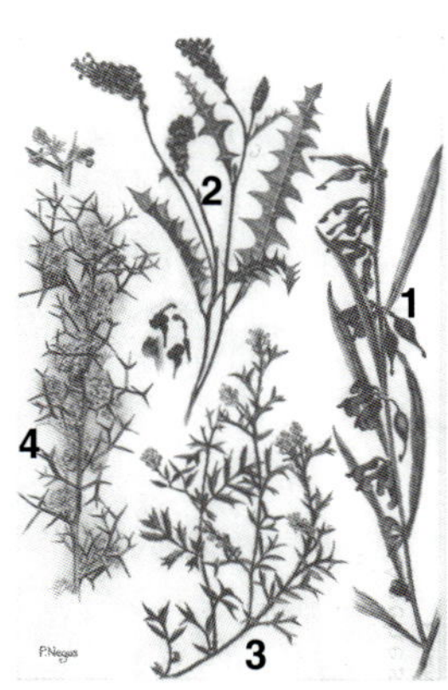

P.Negus

Hakea

After Baron von Hake (1745–1818), German patron of botany.

Flowers of this genus are often very similar to those of *Grevillea*, to which they are closely related. The easiest way to tell them apart is to look at the fruits that, in *Hakea*, develop into hard, woody nuts. These are retained on the shrub, only shedding their seed after a period of drying or after exposure to fire. The seed pods of *Grevillea* flowers, in contrast, fall off when they are ripe. There are over 100 species of *Hakea* in WA, of which I have found nine in the area covered by this book. All except one (*H. ceratophylla*) have sweetly scented, white flowers that are an important source of nectar for insects and birds.

1 ***Hakea amplexicaulis***, Prickly Hakea **05321263**

L. *amplexus* = surrounding + *caulis* = stem; refers to the base of the leaves

A robust shrub with large, prickly leaves, usually on just two or three long branches. These often stand out from the surrounding bush and can be over 2 m high. It is common in Jarrah forest in sand or gravel soils between Perth and Albany.

Flowers: Aug.–Nov.

2 ***Hakea prostrata***, Harsh Hakea **05749980**

L. *prostrata* = prostrate, lying along the ground

This can be confused with *H. amplexicaulis*, but is a much denser shrub with smaller leaves that are broadest at the apex rather than at the base. It was probably named from coastal specimens, as it is indeed a ground-hugging shrub in exposed localities. However, it can become a much more substantial, erect shrub or small tree to 3 or 4 m in sheltered situations. Found in heath and woodlands between Perth and Albany, generally in sandy soils.

Flowers: Aug.–Oct.

3 ***Hakea lissocarpha***, Honeybush **05145511**

G. *lissos* = smooth + *carphe* = dry blade of grass (refers to the bracts surrounding the buds)

As its common name suggests, the flowers of this *Hakea* fill the air with a sweet scent on warm winter days. It is one of the first flowers to appear in the Jarrah forest where it is a common understorey shrub, growing to about 1.5 m. The flowers sometimes have a lovely pinkish tinge. This species is widespread through the south-west between Northampton and Israelite Bay.

Flowers: June–Sept.

4 ***Hakea trifurcata***, Two-leaf Hakea **05474663**

L. *trifurcata* = three-forked (leaves)

This dense shrub grows to 2–3 m and may be up to 5 m across. It is found in thickets and is generally associated in this area with granite soils and outcrops. Interestingly, it has both terete (rounded) and broad, flat leaves that are both found on the same plant. When in flower, its presence is given away by its very strong and distinctive smell. Common from Geraldton to Esperance.

Flowers: July–Oct.

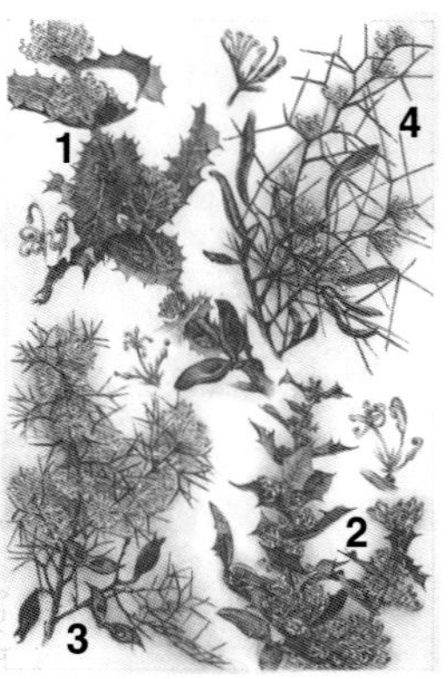

1 *Hakea lasianthoides* **05316308**

G. *lasiantha* = hairy flower + *-oides* = similar to

This tall, open shrub or small tree to 5 m is an elegant member of the Jarrah forest understorey. The flowers are sparse and tend to be hidden in the leaf axils, but young specimens are slender, with attractive reddish hairs along the stems. It grows in sand and clay soils mainly between Busselton and Denmark, with sporadic occurrence north to Perth.

Flowers: Aug.–Nov.

2 *Hakea oleifolia*, Frog Hakea **05749883**

From *Olea* (olive tree) + *folium* = leaf

The scientific name refers to the olive-like leaves of this species, but the common name is much more descriptive, highlighting the fat, warty fruits that, with a little imagination, do appear rather frog-like especially when young and green. This is a rigid shrub or small tree up to 6 m tall, growing in heath or woodland, around the lower west coast from Yallingup to Bremer Bay.

Flowers: Aug.–Oct.

3 *Hakea linearis* **05594936**

L. *linearis* = linear (leaves)

A large, spreading shrub to 3 m, preferring moist situations along creek-lines and winter-wet depressions. It is found around the south coasts from Busselton to east of Albany. The flower clusters, in the leaf axils all along the stems, are very showy at a time when not much else is flowering.

Flowers: Nov.–Feb.

4 *Hakea ruscifolia*, Candle-spike Hakea **05335965**

From *Ruscus,* genus in the Lily family that has small, spiny leaves

Indeed, this is another very prickly one! It is common in sandy or gravelly soils, in heath or Jarrah forest, around the coast from Eneabba to Esperance. It grows to 2–3 m and the flowers often form dense, cylindrical clusters around long stems.

Flowers: Feb.–May

5 *Hakea ceratophylla*, Staghorn Hakea (illustrated below left) **05771854**

G. *ceratos* = horn + *phyllos* = leaf

This 1.5 m shrub is found throughout the south-west from Perth to Albany and along the south coast to Esperance, in heath and woodland. I have only found it in one location here, in a low-lying winter-wet swamp along Forest Grove Road, though I'm sure it occurs elsewhere. In contrast to our other *Hakea* species, its flowers are a lovely russet-brown and green.

Flowers: Sept.–Nov.

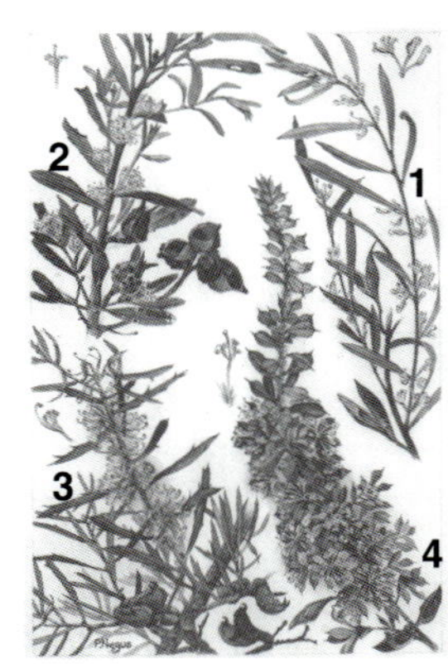

P.Negus

Adenanthos

G. *aden.* = gland + *anthos* = flower; refers to glands at base of ovary

This is a genus of woody shrubs, many having an underground lignotuber that sprouts rapidly after fire. There are around 30 species, only two of which are found outside WA. Unlike the preceding genera, their flowers occur singly along the stems.

1 ***Adenanthos barbiger*** **subsp.** ***intermedius*** ms, Hairy Gland-flower, Hairy Jug-flower **05460832**

L. *barbiger* = bearded

A small, spreading shrub, generally less than 1 m, common in loam or gravel soils in the Jarrah forest between Perth and Augusta. It has a long flowering period, but its brick-red flowers are conspicuous in the winter months when little else is out. This subspecies seems to be intermediate between *A. barbiger* subsp. *barbiger*, which has narrower leaves, and *A. obovatus*. (see below)

Flowers: July–Nov.

2 ***Adenanthos meisneri*** **05594944**

After Carl Meisner (1800–74), professor of medicine and botany at Basle

A dense shrub of sandplain and open woodland, growing to 1.5 m, this shrub dominates the understorey in places. It is common east of the Ridge, in the vicinities of Calgardup and Rainbow Cave Road. The flowers are relatively inconspicuous, but are a lovely deep purple, or, occasionally, cream. Found from Perth to Walpole.

Flowers: much of the year.

3 ***Adenanthos obovatus***, Basket Flower **05145678**

L. *ob* = inverted + *ovatus* = egg-shaped (leaves)

This plant's striking orange-red flowers on long, slender stems, look gorgeous when they catch the sun. It grows to about 1.5 m, in sand or clay soils in heath and woodland and is common between Perth and Albany. This is another early flowerer.

Flowers: May–Sept.

Synaphea

G. *synaphe* = connection; a membrane connects the filament (stem) of the upper sterile anther to the stigma.

These are small shrubs with spikes of yellow, tubular flowers and (usually) divided leaves.

4 ***Synaphea gracillima*** **05145538**

L. *gracillima* = very slender

A low-growing sub-shrub of less than 50 cm, growing on gravelly and sandy soils in woodland and Jarrah forest. This is one of several *Synapheas* that are separated only by small differences in their leaves and flowers. It is found between Moora and Albany.

Flowers: July–Nov.

5 ***Synaphea*** **sp.** ***Redgate Rd. (J. Scott 16)*** **05145627**

Previously thought to be *S. petiolaris* subsp. *triloba*, this has recently been recognised as a separate species and provides an example of the value of ongoing collecting for the WA Herbarium, even in areas that one would think would already have been well-documented. Although small, less than 50 cm high, both flower spikes and leaves make this a showy plant. It grows in clay soil in wet depressions along road verges.

Flowers: July–Sept.

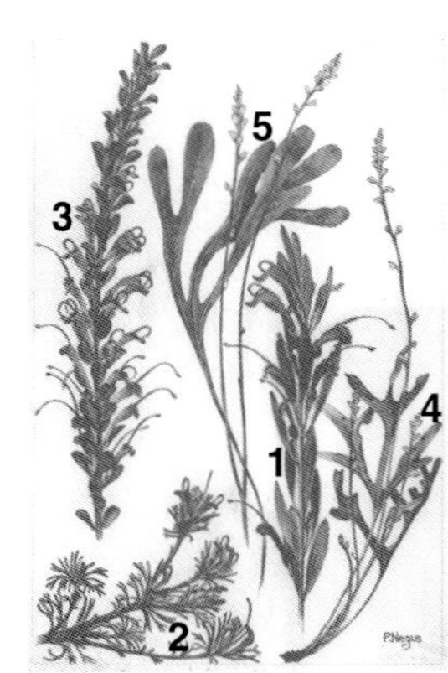

P.Negus

Xylomelon

G. *xylon* = wood + *melon* = apple, fruit

There are two species of this strange genus in WA. Both are small trees with large, heavy woody fruits that remain on the tree until opened by fire or, as in this case, by a period of drying. The large thin 'wing' that helps it waft on the wind away from the parent tree probably aids dispersal of the seed.

1 *Xylomelon occidentale*, Western Woody Pear **05595045**

L. *occidentale* = western

This tree that can grow up to 8 m, looks spectacular when covered with long panicles of white or creamy flowers. It is found between Perth and Augusta, on sandy soils in open woodland or Jarrah forest.

Flowers: Dec.–Feb.

Persoonia

After Christian Persoon (1755–1837), a botanist from the Cape Colony. He studied fungi in Paris.

The two species illustrated are both small trees up to 5 or 6 m. They both have yellow flowers and bright green foliage that stands out amidst the greyer greens of the Jarrah forest where they grow. I have not yet found any other *Persoonias* here, although several species grow inland from Margaret River and along the south coast. Their uncomplimentary common name derives from the fruit that is soft and squishy when very ripe.

2 *Persoonia elliptica*, Spreading Snottygobble **05595037**

G. *ellipticos* = formed like an ellipse (the leaves)

This species has broad, flat leaves and rough grey bark that can become beautifully patterned in older trees. The tree is usually found as scattered individuals on sandy or gravelly soils in Jarrah forest, between the Perth hills, Augusta and Albany.

Flowers: Dec.–Feb.

3 *Persoonia longifolia*, Snottygobble **05595096**

L. *longi* = long + *folium* = leaf

The leaves of common Snottygobble are a give-away, but the bark also has a distinctive flaky texture and is a gorgeous cherry red inside. Note: We urge you not to peel off large chunks of bark, though, as it serves to protect the tree from damage. It has a characteristic, rather crooked shape that would make it look at home in a Japanese garden. Common throughout the Jarrah forest.

Flowers: Dec.–Feb.

Isopogon

G. *isos* = equal + *pogon* = beard; refers to the nut, which is hairy all over

Isopogons have flowers in heads, but unlike the very similar *Petrophile* (overleaf), the cone-scales fall off with the nuts.

4 *Isopogon sphaerocephalus*, Drumstick Isopogon **05626668**

(illustrated below left)

G. sphaira = sphere + cephale = head

This is a stiffly erect, hairy shrub to about 1 m with pale yellow flowers. It grows on gravel soil and is found through the Jarrah forest between Perth and Albany. It mainly grows to the east of our area.

Flowers: Oct.–Nov.

4 *Isopogon sphaerocephalus*, Drumstick Isopogon

P.Negus

Petrophile

G. *petros* = rock + *phileo* = to love

Many *Petrophile* species are found in rocky habitats, though most are quite happy elsewhere. They have heads of flowers, with their fruits remaining in a cone of persistent bracts.

1 *Petrophile linearis*, Pixie Mops 05772214

L. *linearis* = linear (leaves)

A small shrub less than 1 m, but with spectacular pink flowers, growing on sand in Banksia or Jarrah woodland. Its distribution is all along the coastal plain from the Hill River to Augusta.

Flowers: Sept.–Nov.

2 *Petrophile serruiae* 05771838

From the genus *Serruia*, which has similar leaves

A very prickly shrub to 1.5 m high. Here it grows in limestone sands along the Leeuwin–Naturaliste Ridge and its flowers are pink. It also grows on the lateritic soils of the Darling Range from Geraldton to Albany, but in this habitat the flowers are yellow.

Flowers: Aug.–Nov.

3 *Petrophile diversifolia* 05772184

The species name refers to its leaves, which differ markedly between young and old plants.

This species is common in patches through the Jarrah forest. It likes sandy or gravelly soils and is found between Margaret River and Albany. It can grow more than 2 m tall on single, slender stems and has white flowers.

Flowers: Oct.–Nov.

Conospermum

G. *conos* = cone + *spermum* = seed (having cone-shaped nuts)

Many members of this genus are commonly called smokebushes, because of the likeness of their grey masses of woolly flower heads to drifts of smoke. The species found here, however, are atypical in this regard, having more or less glabrous (hairless) flowers that bear little resemblance to smoke.

4 *Conospermum paniculatum*, Tangled Smokebush △ 05460964

L. *panicula* = tuft, panicle of flowers

This is an open, spreading shrub up to 1 m high with white or very pale blue flowers borne on long, sinuous stems. The long, linear leaves arise from the plant base. It grows in sandy or clayey soils and its main distribution is to the east and north of our area, between Busselton, Augusta and Nannup.

Flowers: Aug.–Nov.

The very similar*C. flexuosum* is also found in the area. It has more zig-zag (flexuose) flower stems.

5 *Conospermum capitatum* subsp. *glabratum* (illustrated below) 05749972

L. *capitatum* = capitate, forming a head

This peculiar, wiry little shrub, only 50 cm high, is distinctive, but not easy to see in the tall heath or low Banksia woodland in which it grows. Its flowers are almost indiscernible, hidden amongst the masses of its long, curly leaves and it is often mistaken for a *Petrophile*. It likes sandy soils and is found mostly between Busselton and Albany.

Flowers: Sept.–Oct.

5 *Conospermum capitatum* subsp. ***glabratum***

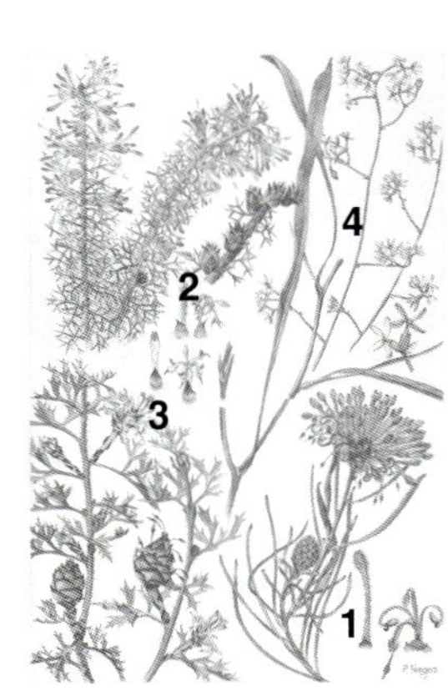

P. Negus

Loranthaceae

This is a family of semi-parasitic shrubs and trees that grow attached to the branches or, in the case of *Nuytsia*, to the roots of their hosts from which they obtain water and minerals. They do still photosynthesise, producing carbohydrates as do normal plants. The family includes the Australian mistletoes, *Lysiana* and *Amyema*. We have not found members of either of these genera in the Augusta–Margaret River area.

Nuytsia

After Pieter Nuyts (?1600–?33), of the council of the Dutch East Indies, who sailed with the *Gulde Zeepard* in 1627 along the south and west coasts of Australia.

1 *Nuytsia floribunda*, West Australian Christmas Tree

L. *floris* = flower + *abundus* = abounding in

Most people will be familiar with this spectacular tree. It grows to about 7 m on sandy or granitic soils in forest or heathland and is found from Kalbarri to Israelite Bay. It has a dry, papery, three-winged fruit. It is parasitic on the roots of many species, often connected by runners to hosts some distance away. It has been known to attack underground telephone cables and plastic reticulation tubing in a case of mistaken identity!

Flowers: Dec.–Jan.

Lauraceae

This family that includes the laurel and cinnamon trees, is represented in south-west WA by just one genus, *Cassytha*: from the Arabic, *kasutha,* a plant resembling dodder. There are about 11 species currently recognised.

Note: One of the true dodders, *Cuscuta epithymum,* a weed, is probably also in the area. It belongs to a different family, Cuscutaceae.

2 *Cassytha racemosa*, Dodder Laurel **05577454**

L. *racemosa* = bunch of grapes; flower heads in which the terminal flowers are the last to develop.

A parasitic perennial herb and climber that eventually smothers its host completely. It saps nutrients by means of haustoria, specialised roots that can penetrate the host's stems. Its leaves are reduced to scales and it has masses of tiny white flowers. Found right down the west of the state from Cape Range and along the south coast to the Great Australian Bight.

Flowers: much of the year.

Note: A very similar species, but with yellow flowers, *Cassytha flava,* is also found here.

Orobanchaceae

Orobanche: the name used by Theophrastus for Dodder, *Cuscuta europaea,* and by Pliny for species of *Orobanche*!

3 **Orobanche minor*, Lesser Broomrape **05750121**

L. *minor* = smaller

This is an introduced species that is native to Europe. It grows to about 40 cm. Broomrapes are truly parasitic (on roots), having no chlorophyll. Roots grow out from the bulb-like structure at its base and penetrate those of the host. You will find it growing in clay, sand or gravel on dunes, roadsides, granite outcrops, and gardens. It has a scattered distribution throughout the south-west from Shark Bay to Ravensthorpe.

Flowers: Sept.–Dec.

Santalaceae

This is a worldwide family of herbs, shrubs and trees, with about 50 species in Australia. Many species are semi-parasitic, taking nutrients from the roots of other plants, but they also contain chlorophyll with which they manufacture their own food as do normal plants.

Exocarpos

G. *Exo* = outside + *carpos* = fruit; refers to the pedicel (stalk) of the fruit that is fleshy like a fruit itself.

4 *Exocarpos sparteus*, Broom Ballart, Native Cherry **05595010**

L. *spartum* = esparto, Spanish broom

A weeping shrub that, in heathland, is pruned to the general height of the vegetation, but in forest can reach 4 m. It is yellowish-green, with tiny yellow flower spikes, but more prominent, reddish berries appearing in late spring and summer. On close inspection, the older stems often have interesting striated patterns. This species grows on coastal limestone here, but is widely distributed and is found also in inland sandplain and desert habitats.

Flowers: Feb.–Oct.

Leptomeria

G. *Leptos* = slender + *meria* = part (referring to the branchlets)

5 *Leptomeria squarrulosa* **05145554**

L. *squarrulosus* = having small, overlapping scales

An erect, spreading shrub to about 1 m, with tiny leaves and flowers, looking rather like a miniature pine tree. It grows in the Jarrah forests of the lower south-west, between the Capes and Albany, on sandy or lateritic soils.

Flowers: June–Oct.

Santalum

G. *santalon,* from the Arabic *sandal,* Indian sandalwood

6 *Santalum acuminatum*, Quandong (illustrated page 76) **05795303**

L. *acuminatum* = sharp-pointed, drawn into a long, narrow point

There are several populations of quandongs growing along the cliff-tops of the Leeuwin–Naturaliste Ridge. This plant is widespread further north along the coast and inland throughout the arid region, where it may be a shrub or small tree up to 7 m tall. On the windswept coastal heath where it is found here it is generally a stunted small shrub less than 1 m tall. So far, I have found plants only with flowers and have not yet found any fruiting.

Flowers: Dec.–Mar.

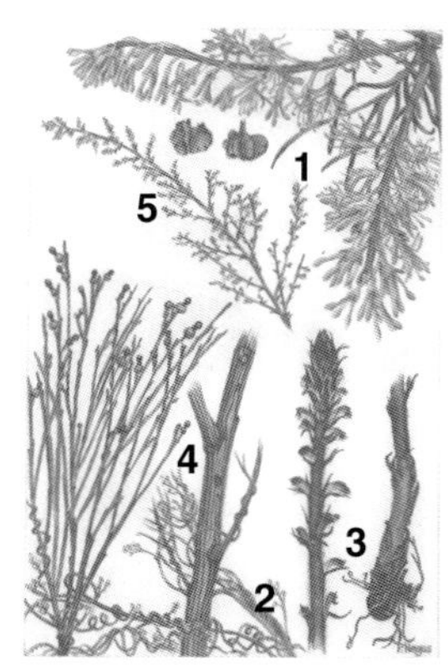

P. Negus

Polygonaceae

Members of this family include the docks, sorrels and the infamous double-gees (see p. 200), all of which are weeds in WA, and we have our share of them in the Augusta–Margaret River area! The species below, however, are native to the area.

Muehlenbeckia

After Heinrich Muehlenbeck (1798–1845), a physician and botanist from Alsace, in France.

1 *Muehlenbeckia adpressa*, Climbing Lignum **05772451**

L. *ad* = towards + *pressus* = pressed, flattened (we are not sure what!)

Climbing Lignum is very prevalent in developed areas along the coast, where it tends to run rampant in cleared and disturbed areas, strangling other vegetation and you may be forgiven for mistaking it for a weed. But it is a native plant, common in south-west coastal heath and throughout the drier inland. In undisturbed bushland it does not seem to be nearly as vigorous, presumably due to competition from other species.

Flowers: Sept.–Dec.

Chenopodiaceae

The other plants illustrated on this page are from this large family. Many of its members are characterised by having fleshy stems or leaves and they are often found in salty habitats, either close to the sea, or in the arid interior. Their flowers are generally small and inconspicuous and are probably wind pollinated.

Rhagodia

G. *rhagodes* = resembling berries or grapes

2 *Rhagodia baccata* subsp. *baccata*, Berry Saltbush **05595029**

L. *baca* = berry, having fleshy or pulpy berries

This is a common shrub of sand-dunes and limestone that grows in floppy, tangled fashion to 2 m. Its form varies considerably from a low, dense bush with thick fleshy leaves near the sea, to a straggly, tall, spindly shrub in shaded woodland. It has striking dark red berries in autumn, well-loved by silver-eyes. If their ripening coincides with that of the grapes, it can be a blessing for vignerons! It is found all around the south-west and southern coasts. Flowers: Nov.–Jan.

Atriplex

Name used by Pliny for mountain spinach, an *Atriplex* species. Plants of the genus have bracteoles that enlarge after flowering and cover the fruit.

3 *Atriplex isatidea*, Coastal Saltbush **05456754**

G. *Isatis* = name of a plant that produced woad, a blue dye

An erect shrub that can grow to 2.5 m, though it is often much lower. It has lovely silvery foliage and is found in sand all along the WA coast.

Flowers: mainly Mar.–June.

Threlkeldia

After Caleb Threlkeld (1676–1728), an English botanist who published an account of the Irish flora.

4 *Threlkeldia diffusa*, Coast Bonefruit, Wallaby Saltbush **05456835**

L. *diffusa* = loosely spreading

A prostrate, spreading perennial with succulent leaves. Widely distributed on coastal limestone and, much less commonly, on inland saline flats. The tiny greenish flowers are difficult to see and the fruits become apparent only when they ripen from green to dark purple.

Flowers: Oct.–Nov.

6 *Santalum acuminatum*,
Quandong (described page 74)

Sarcocornia

G. *Sarcos* = fleshy + *cornu* = horn; refers to the horn-shaped branches

5 *Sarcocornia blackiana*, Glasswort **05772427**

After John McConnell Black (1855–1951), author of Flora of South Australia

A semi-erect, shrubby perennial herb, not much more than 50 cm high, with succulent, jointed stems that send down roots at the nodes. It is common at the edge of vegetation along the coast, wherever there are saline seepages or waterlogged areas and it can grow on very shallow soil over granite or limestone. The blue-green stems redden with age, but you need to look carefully to see the tiny yellow flowers. Found from Carnarvon to the Great Australian Bight.

Flowers: Nov.–Feb.

P.Negus

Portulacaceae

Calandrinia

After Jean Luis Calandrini (1703–58), professor of mathematics and botany at Geneva

This is a small group of succulent herbs, mainly found around salt-lakes and sand-flats of the drier inland.

1 *Calandrinia liniflora* (illustrated below) **05626625**

L. *linum* = flax + *flos, floris* = flower

An attractive, slender little annual herb to about 30 cm, with several stems bearing pink flowers and a basal tuft of succulent, linear leaves. It grows on the sandy soils of the Leeuwin–Naturaliste Ridge and is found around the coast from Kalbarri to Albany.

Flowers: Oct.–Nov.

Aizoaceae

G. *aiezoos* = evergreen, everlasting

This is a large cosmopolitan family of creeping, succulent, fleshy-leaved plants adapted to arid or saline habitats, both inland and on the coast.

Carpobrotus

G. *carpos* = fruit + *brotos* = edible

2 Carpobrotus virescens, Coastal Pigface

L. *virescens* = becoming green or verdant, light green

This is the species of Pigface common all along the dunes of the Capes coast, whereas *C. edulis*, with larger, yellow flowers, is the one people will be more familiar with around Perth. The bright purple flowers and red fruits of *C. virescens* add colour to the coast throughout the summer months. Though maybe not as sweet as the introduced *C. edulis*, its fruits, sometimes known as 'bush strawberries', are quite pleasant when ripe. *Carpobrotus* is also a valuable dune stabiliser. It has a patchy distribution right around the coastline from Shark Bay to the Great Australian Bight.

Flowers: Sept.–Jan.

Tetragonia

G. *Tetra* = four + *gonia* = corner, angle (refers to the shape of the fruit of some species)

Both species here have yellow flowers.

3 *Tetragonia decumbens, Sea Spinach **05611865**

L. *decumbens* = reclining, prostrate with the tips pointing upwards

A decumbent or trailing low shrub, less than 50 cm high, common on sand-dunes. Its extensive root system, often several metres long, makes it a valuable dune stabiliser. It is native to South Africa, but has become naturalised here and is found around the coast from Geraldton to Bremer Bay.

Flowers: April–Dec.

4 *Tetragonia implexicoma*, Bower Spinach **05456819**

L. *implexus* = interwoven, entwined + *coma* = hair

A prostrate, scrambling perennial, more slender than *T. decumbens*. The fruit is also different, being globular and succulent, rather than leathery and winged. It is also found on the coastal dunes in certain areas right around the west and south coasts.

Flowers: mainly Aug.–Oct.

Brassicaceae

This is a huge, worldwide family including many well-known vegetables and agricultural crops–cabbage, turnip, mustard, rapeseed, to name a few. Many species found in WA are introduced, although we have a fair number of indigenous species also.

5 *Cakile maritima, Sea Rocket **05594987**

L. *maritimus* = growing by the sea

This succulent, annual herb is a native of Europe and Africa, but is naturalised in Western Australia and is found around the coast from Shark Bay to Esperance. It is a coloniser, one of the first plants to gain a foothold on the mobile fore-dunes behind the beach. The flowers are pale mauve.

Flowers: all year.

Stenopetalum

G. *stenos* = narrow + *petalum* = petal

6 Stenopetalum robustum **05460816**

L. *robustum* = hard, firm

A small, dainty annual with white flowers, found in heathland on the sandy soils of the coastal limestone. It is probably widely distributed around the south-west coast, though it has mostly been collected along the south coast between the Capes and Albany.

Flowers: Sept.–Oct.

1 *Calandrinia liniflora*

P.Negus

Amaranthaceae

Amaranthus means 'unfading flower' and many members of the Amaranth family are showy and colourful. We have found just two species of *Ptilotus* here, both small herbs of the coastal heath.

Ptilotus, the Mulla-mullas

G. *ptilotos* = winged, maybe feathered; many species have long-haired (feathery) sepals

Ptilotus has about 100 species, only one of which is found outside Australia. Most are found in the arid interior.

1 ***Ptilotus manglesii***, Rose-tipped Mulla-mulla, Pom-poms **05772125**

After Captain James Mangles (1786–1867), naturalist and patron of botany

Although small, this little perennial herb is a colourful one, each of the tiny flowers on its dense spike having bright pink petals amongst long woolly hairs. It is prostrate, but with upturned stems to 60 cm. Widespread between Kalbarri and Bremer Bay.

Flowers: Nov.–Dec.

2 ***Ptilotus drummondii***, Narrow-leaf Mulla-mulla, Pussytail **05577306**

After James Drummond (1784–1863), who accompanied Captain Stirling to WA on the *Parmelia* in 1829 and, although never officially appointed or paid, was Western Australia's first Government Horticulturalist.

A less conspicuous species, this is a straggly perennial herb to about 80 cm with greenish pink pom-pom flower heads. Found throughout the state.

Flowers: Sept.–Nov.

Ranunculaceae

Ranunculus, Buttercups

Ranunculus means "little frog" because many species live in marshy areas. Two species are shown here, one native and the other introduced, but there may be more species to be found in the area.

3 ***Ranunculus colonorum***, Common Buttercup **05515394**

L. *colonorum* = of the colonists (this plant was found in the early days of the colony of WA in 1833).

A slender, erect, perennial herb to about 80 cm. This plant was probably recognised early because of its familiarity, buttercups being well-known from the northern hemisphere. This one is found on sandy or clay soils in Jarrah or Karri forest, often in damp situations, between Perth and Albany.

Flowers: Sept.–Dec.

4 ***Ranunculus muricatus***, Sharp Buttercup **05515343**

L. *murex* = a fish with a rough shell; muricate–rough, with short, hard points

This is a usually prostrate and slightly succulent annual herb, less than 30 cm high. It prefers peaty sand in winter-wet depressions, swamps, or running water and is occasionally found as a garden weed. Found around the coast from the Moore River to Albany. It is native to the Mediterranean.

Flowers: Aug.–Dec.

Clematis

G. from *clema* = vine shoot, on account of its climbing habit

8 ***Clematis pubescens***, Common Clematis, Old Man's Beard (illustrated page 83) **05474426**

L. *pubes* = soft, downy hair

This lovely plant is one of our most common climbers, adorning the forest understorey with a mass of white blooms in early spring and silvery-grey, woolly seed-heads later on. It is usually found on sandy soils and is widespread on the coastal plain and forest between Perth and Israelite Bay.

Flowers: Aug.–Oct.

P.Negus

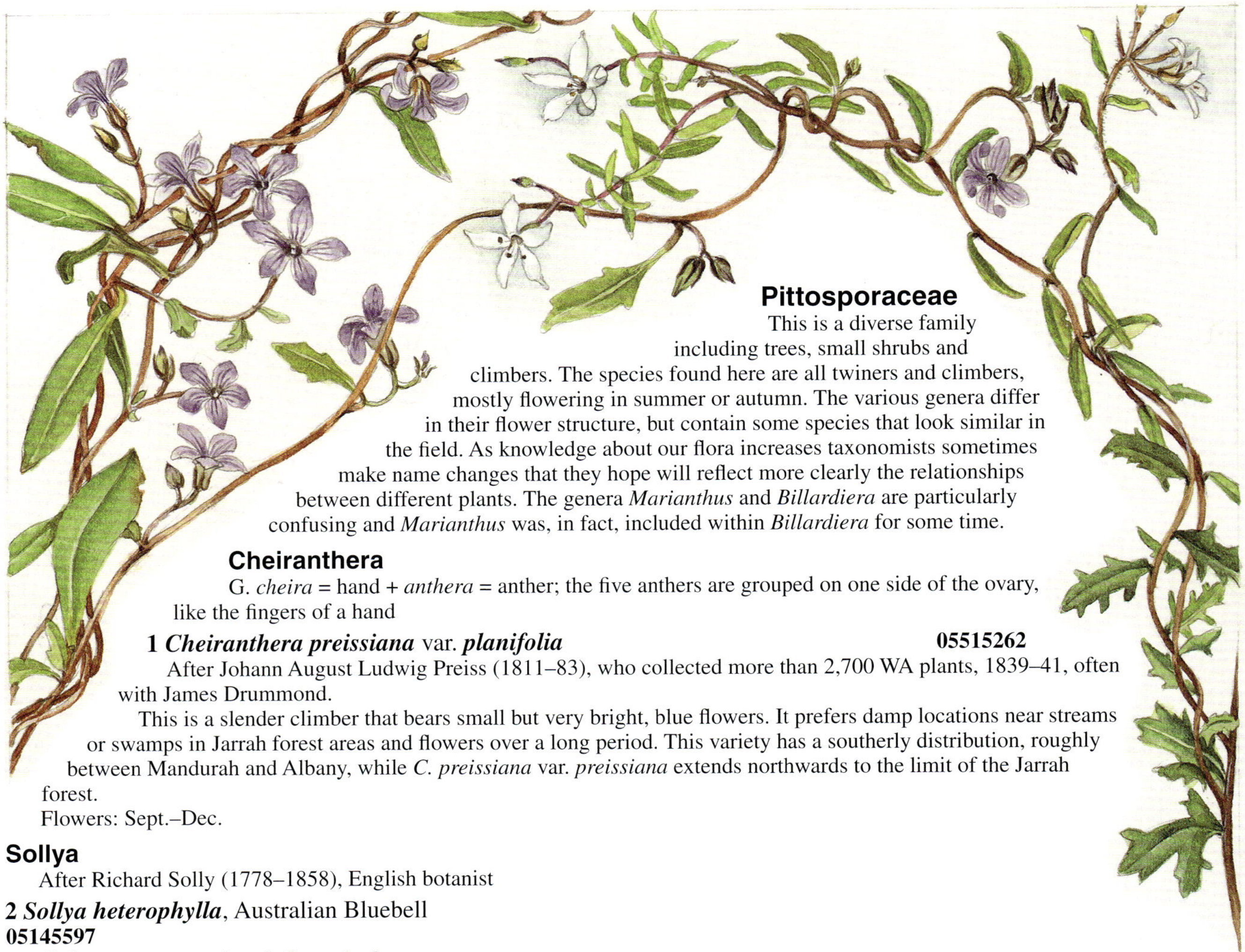

Pittosporaceae

This is a diverse family including trees, small shrubs and climbers. The species found here are all twiners and climbers, mostly flowering in summer or autumn. The various genera differ in their flower structure, but contain some species that look similar in the field. As knowledge about our flora increases taxonomists sometimes make name changes that they hope will reflect more clearly the relationships between different plants. The genera *Marianthus* and *Billardiera* are particularly confusing and *Marianthus* was, in fact, included within *Billardiera* for some time.

Cheiranthera

G. *cheira* = hand + *anthera* = anther; the five anthers are grouped on one side of the ovary, like the fingers of a hand

1 *Cheiranthera preissiana* var. ***planifolia*** **05515262**

After Johann August Ludwig Preiss (1811–83), who collected more than 2,700 WA plants, 1839–41, often with James Drummond.

This is a slender climber that bears small but very bright, blue flowers. It prefers damp locations near streams or swamps in Jarrah forest areas and flowers over a long period. This variety has a southerly distribution, roughly between Mandurah and Albany, while *C. preissiana* var. *preissiana* extends northwards to the limit of the Jarrah forest.

Flowers: Sept.–Dec.

Sollya

After Richard Solly (1778–1858), English botanist

2 *Sollya heterophylla*, Australian Bluebell
05145597

G. *heteros* = unequal + *phyllon* = leaf

A sturdy, shrubby twiner that can grow to over 2 m. It is a well-known species in cultivation and occurs in the wild throughout the south-west, on a variety of soils. It grows in open woodlands and heath. In some parts of the Eastern States this plant has become a weed.

Flowers: all year, but mainly Nov.–Feb.

Marianthus

Maria, Virgin Mary (because of the white flowers of the original species, *M. candidus*).

3 *Marianthus candidus* **05515467**

G. *candidus* = white

A shrubby climber found here in heathland, on the sandy soils of the limestone ridge. The creamy white flowers are small, but showy, growing in dense corymbs. It is found from the Perth hills south to Albany.

Flowers: Oct.–Jan.

4 *Marianthus tenuis* (illustrated above left) **05594871**

L. *tenuis* = thin, narrow, delicate

M. tenuis is a slender little climber, covering bushes in the Jarrah forest through late summer and autumn with its pale blue flowers. Its light shade of blue and dark spots distinguishes it from *Billardiera variifolia* that also flowers earlier. It is mainly found from the Capes around the south coast to Albany.

Flowers: Mar.–May

Billardiera

After Jacques Julien Houton de le Billardiere (1755–1834), French botanist on d'Entrecasteaux's voyage to WA and Tasmania in 1791–94.

5 *Billardiera floribunda* **05595088**

L. *floris* = flower + *abundus* = abounding in

Its robust nature and large white flowers distinguish this climber from the somewhat similar *Marianthus candidus*. It also has a different habitat, growing on sand or laterite in Jarrah forest. It is one of our most spectacular summer wildflowers and is found along the Darling Range and through the lower south-west to Albany.

Flowers: Dec.–Feb.

6 *Billardiera laxiflora* (illustrated above right) **05594898**

L. *laxus* = wide, loose, open + *floris* = flower

This is another slender twiner, sparsely flowered and less obvious than *B. floribunda*. The difference between its young, toothed leaves and older, linear ones is quite noticeable. It occurs in Jarrah forest, mainly between Bunbury and Albany.

Flowers: Feb.–April

♂
P.Negus

7 *Billardiera variifolia* (illustrated previous page) **05335914**

L. *varius* = changeable + *folium* = leaf

This species has deep blue to purple flowers and is again a climber of the Jarrah forest, growing on sand, loam or clay soils, often over laterite. It tends to be more robust than *Marianthus tenuis* (see above) and flowers earlier, though they are sometimes found together. Again found between Perth and Albany.

Flowers: Dec.–Mar.

Dilleniaceae

This mainly tropical family is represented in the south-west of WA by just one genus, *Hibbertia*.

Hibbertia, Guinea Flowers

After George Hibbert (1757–1837), a nurseryman of Clapham, London, whose herbarium was presented to the Linnean Society.

There are nearly 90 species of *Hibbertia* in WA, most of which are found in the south-west. These are all yellow-flowered, apart from three bright orange-coloured species. Because their flowers are similar, people often do not realise how many different species there are. I have so far found 10 in our area.

1 *Hibbertia cuneiformis*, Cutleaf Hibbertia **05145686**

L. *cuneus* = wedge + *forma* = shape

One of our most spectacular shrubs, *H. cuneiformis* is common on sand or loam soils in the Karri forest, but also on coastal dunes. It is often wind-pruned to a low, sprawling bush on the coast, but grows to be a substantial shrub up to 3 m high in the shelter of the forest. Individual, old leaves turn a beautiful red and the clusters of bright orange fruits are also attractive. Found around the south-west corner, between Rockingham and Albany and along the south coast to Esperance.

Flowers: June–Nov.

2 *Hibbertia grossulariifolia* **05460972**

From the genus Grossularia + L. *folium* = leaf

Look for the soft, hairy leaves of this species on the Karri forest floor, or in the shelter of dense shrubs amongst the sand dunes where they are protected from desiccation. It has large showy flowers that start to appear in early spring. It is a prostrate, woody perennial, growing vigorously on sand or loam soils over limestone. It is just found in the south-west corner from Cape Naturaliste to Albany.

Flowers: Aug.–Nov.

3 *Hibbertia furfuracea* **05474620**

L. *furfur* = scale, + *-aceous* (suffix), scaly, scurfy

This species can be confused with *H. hypericoides* (below), but it has longer leaves and is a taller, more open shrub, growing up to 2 m. It is less widely distributed, but is common in our area on sandy soils on the eastern, sheltered side of the limestone ridge. It occurs in near coastal locations and in Karri forest from the Capes to Albany.

Flowers: Aug.–Oct.

4 *Hibbertia hypericoides*, Yellow Buttercups **05145473**

From the genus Hypericum + *-oides* = like

This is probably the best known *Hibbertia* of all, being one of the most common understorey shrubs of the Jarrah forest and the coastal plain, between Geraldton and Augusta. It is a small woody shrub, growing to about 1 m. After fires, its new growth is often a multitude of erect stems that become branched and bushier with age. It has a long flowering period.

Flowers: June–Nov.

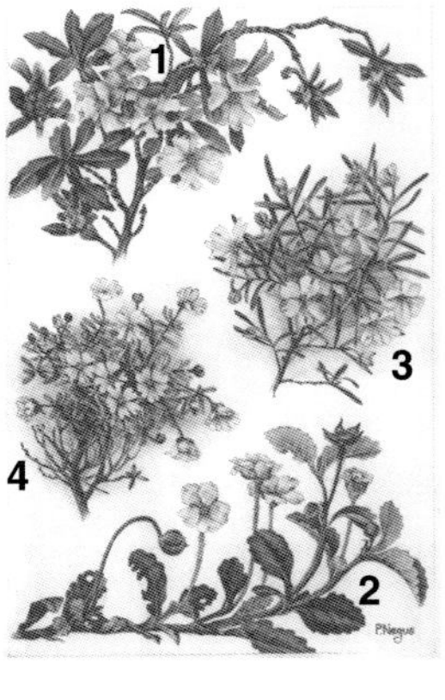

P.Negus

1 *Hibbertia cunninghamii* **05316324**

After Allan Cunningham (1791–1839), early botanical collector, mainly in NSW.

A slender, sprawling sub-shrub to about 50 cm high, growing on sandy or loamy soils over limestone or laterite in coastal woodlands or Jarrah forest. It has long, stem-clasping leaves and can be confused with the very closely related *H. amplexicaulis*, in which the leaves are broader and the flowers larger. There are many intermediate forms and these two species may be classed together in the future. They are both common between Perth and Albany.

Flowers: Aug.–Nov.

2 *Hibbertia inconspicua* **05577209**

L. *inconspicuus* = undistinguished

A prostrate or low sub-shrub growing only to about 50 cm. It is found on sandy, clay or lateritic soils in Jarrah forest. As the name indicates, it is not a particularly showy plant, but its grey-green foliage and mass of small, stalked flowers are nevertheless attractive. Another species of the lower south-west, found between the Capes and Albany.

Flowers: Oct.–Nov.

3 *Hibbertia pilosa*, Hairy Guinea Flower **05474639**

L. *pilosus* = hairy

This species is an erect, often single stemmed, or spreading shrub, to about 1 m. It is one of several very closely related species, the names of which are currently being revised. It is actually quite a smooth plant, but its name refers to its very silky, hairy sepals. It has broader leaves than *H. hypericoides*, but they are not stem-clasping like *H. cunninghamii*. To distinguish it look for the shiny, brown bracts surrounding each flower. It grows on sandy or lateritic soils in Jarrah forest or on the coastal sandplain from Perth south to Augusta and around to Albany.

Flowers: Sept.–Oct.

4 *Hibbertia racemosa*, Stalked Guinea Flower **05460859**

L. *racemus* = bunch of grapes, head of flowers with terminal ones being the last to develop

A prostrate, spreading or ascending shrub, usually less than 50 cm, with small flowers on long stalks. It likes to grow on the sandy, limestone soils of coastal areas and occurs from Shark Bay right around the west and south coasts to Esperance.

Flowers: Oct.–Dec.

5 *Hibbertia notibractea* ms [formerly ***H.*** sp. ***"rigid bracts" (J.R. Wheeler 3220)***] **05461030**

Found in sandy areas of Jarrah and coastal woodlands between the Capes and Albany, this species is quite common, but has only recently been given a specific name. It is a small, spreading shrub less than 50 cm high. It looks rather like *H. racemosa* and grows in the same type of location, but its flowers are not stalked and its leaves are shorter.

Flowers: Sept.–Nov.

6 *Hibbertia rhadinopoda* **05771897**

G. *rhadinos* = slender + *podos* = foot; refers to the slender flower stalks, or pedicels

This prostrate, mat-forming little ground cover adorns the damp soil patches on granite outcrops, as well as creek banks and swampy areas with its mass of small yellow flowers. It prefers sandy and lateritic soils and ranges from Bindoon to Albany.

Flowers: Oct.–Dec.

Frankeniaceae

This is a family of low herbs or sub-shrubs with opposite, often crowded leaves, growing in saline, arid or coastal areas.

Frankenia

After Johann Franke (1590–1661), professor of botany at Uppsala, who first enumerated the plants of Sweden.

7 *Frankenia pauciflora*, Sea Heath **05595053**

L. *paucus* = few + *floris* = flower

A twiggy, little shrub up to 50 cm high, with branches becoming gnarled with age. It nestles in amongst the boulders and rocks in the spray zone. This hardy plant is widespread around the WA coast and is also found on saline flats and inland salt lakes.

Flowers: Dec.–Feb.

6 *Hibbertia rhadinopoda*

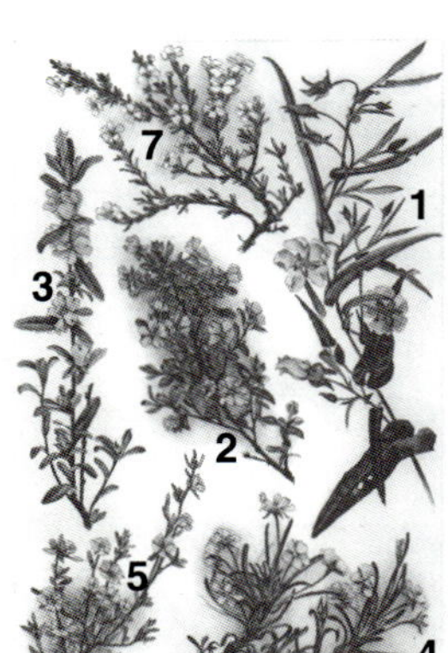

P.Negus

Droseraceae

This is a fascinating worldwide family of plants that reach their greatest diversity in the south-west of WA. They supplement their nutrition by trapping and ingesting insects with the aid of sticky, glandular-hairy leaves that secrete enzymes to digest their captured prey. They generally grow in places where the soils are deficient in essential nutrients and they are well-represented here. They are usually perennial herbs with a wide variety of forms and are often tuberous.

Drosera, sundews

G. *droseros* = dewy

1 *Drosera glanduligera*, Pimpernel Sundew **05749999**

L. *glandulae* = glands + *gero* = to bear

A tiny, but very common sundew, only a few centimetres tall, of open Jarrah forest or heath, generally in moist situations. This one is an annual. It is found from Kalbarri to Esperance.

Flowers: Sept.–Nov.

2 *Drosera macrantha* subsp. ***macrantha***, Bridal Rainbow **05461006**

G. *macros* = large + *anthos* = flower

This is a climber with large, pale pink flowers, delicate stems and prominent, orbicular leaves that are of two types, short and long stalked. Here, it is found on sandy soils, mainly in coastal heath and woodland. It has a widespread distribution from Shark Bay to Esperance.

Flowers: Aug.–Oct.

3 *Drosera marchantii* subsp. ***marchantii*** △ **05624142**

After Dr. Neville Marchant, director of the Western Australian Herbarium from 1991

This species is a dainty, erect perennial with dark pink flowers, growing to about 40 cm. This plant apparently prefers damp locations, though the location of the collected specimen was in dry Jarrah forest. Mostly found between Mandurah, Collie and Busselton.

Flowers: Aug.–Oct.

4 *Drosera pallida*, Pale Rainbow **05460999**

L. *pallidus* = pale

Another climber, but with loose heads of white flowers and small reddish leaves on long pedicels. Again found on sandy soils in heath, woodland or Jarrah forest, between Geraldton and Albany.

Flowers: Aug.–Oct.

5 *Drosera erythrorhiza*, Red-ink Sundew

G. *erythros* = red + *rhiza* = root

This identification is tentative, as I have not yet found flowers. These are white, appearing in autumn before the leaves develop. The rosette of leaves is very evident through winter, but withers away in spring, leaving an underground tuber that is bright red. This little plant is common, growing in large colonies on sandy soil, in woodland and open forest, between Three Springs and Ongerup.

Flowers: May–June

6 *Drosera menziesii* subsp. ***menziesii***, Pink Rainbow (illustrated page 7) **05625270**

After Archibald Menzies (1754–1842), surgeon, who visited King George Sound and collected with Vancouver's voyage, 1790–95.

A delicate, twining sundew with red stems and leaves, and large, usually deep pink flowers. It grows in a variety of habitats, but is often found here on granite outcrops. Widely occurring between Shark Bay and Cape Arid.

Flowers: Sept.–Oct.

7 *Drosera sulphurea* (illustrated page 8) **05626854**

L. *sulphureus* = pale, clear yellow

A beautiful erect, twining perennial up to 60 cm tall. Its bright yellow flowers are conspicuous amongst the sandplain vegetation in which it grows. Found here on the Nindup Plain.

Flowers: Dec.–Jan.

8 *Drosera stelliflora*, Pygmy Sundew **05624827**

L. *stelliger* = starry + *flos, floris* = flower

This is one of several tiny sundews that are found in the south-west. It grows in colonies on sandy heathland and is only about 1 cm high. You can easily see its clumped red leaves, but have to look very hard for the minute flowers. As with several pygmy sundews, the basal leaves persist when they die off, causing the plant to grow from the top of a small mound of dead leaves and sand. This species has a localised distribution in the Busselton and Capes area.

Flowers: Oct.–Nov.

9 *Drosera pulchella*, Pretty Sundew **05626846**

L. *pulchellus* = beautiful and small

Another pygmy sundew only a few mm high, with pink, white or orange flowers. Its rosette of basal leaves with their flattened stems is distinctive. It grows on peaty sand in swampy areas and has a near coastal distribution between Bindoon and Albany.

Flowers: Nov.–Jan.

Lentibulariaceae

This is another group of insectivorous plants and although the family is not closely related to the Droseraceae, it seemed logical to place them together. They are small herbs growing in waterlogged and swampy habitats that are generally poor in nutrients.

Utricularia, Bladderworts

L. *utriculus* = small bladder, or leather bottle

Plants have tiny bladders attached along their basal stems that are submerged in mud or water. These have intricate systems of valves and bristles that can trap and ingest minute water-borne creatures.

10 *Utricularia multifida*, Pink Petticoats **05771943**

L. *multus* = many + *fidus* = cleft, divided into many parts

A tiny herb, but prolific amongst the moss-mats of winter-wet depressions and granite outcrops. When flowering en-masse they make a beautiful display. Found from Jurien Bay to Albany.

Flowers: Sept.–Nov.

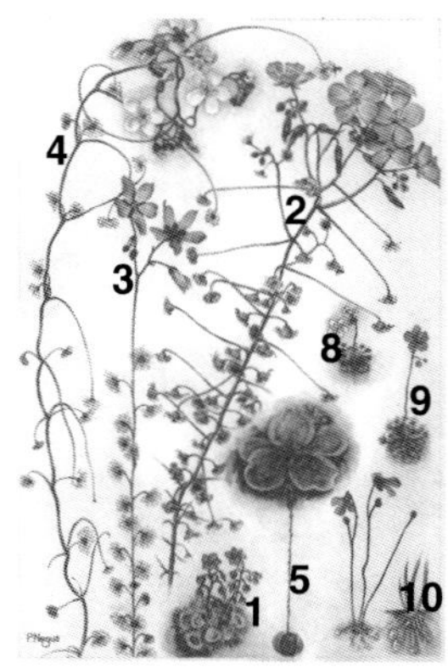

P.Negus

Mimosaceae

This is Australia's largest plant family. Its members are legumes that have root nodules in which they house bacteria that extract nitrogen from the air and convert it to nitrate compounds that enhance the plant's growth. Mimosaceae can thus thrive in the country's poorest soils. The largest and most well-known genus is *Acacia*, with well over 400 recognised species in the south-west of WA alone.

Paraserianthes (formerly Albizia)

1 ***Paraserianthes lophantha*** var. ***lophantha***, Cape Leeuwin Wattle **05626951**

G. *lophos* = crest on helmet or bird + *anthos* = flower

A tall shrub or tree to 10 m with bipinnate leaves and large, cylindrical spikes of greenish-yellow flowers. The long stamens, united in a tube, separate this genus from *Acacia*. Cape Leeuwin Wattle is a common understorey tree in the Karri forest though with a wider distribution between Perth and Esperance. Like many of the legumes it regenerates prolifically after fire and colonises disturbed areas.

Flowers: July–Sept.

Acacia, Wattles

G. *acis* = a pointed instrument; many exotic species have sharp prickles. Though only a few WA species have spines, several local ones have pungent (sharp pointed) phyllodes.

Acacia species range from tall trees to low sub-shrubs and are found in most habitats, though they are more prevalent in drier areas. The adult plants of most species do not have true leaves. Instead, flattened leaf stems called phyllodes replace the leaves. Phyllodes lose water from evaporation less quickly than true leaves and are therefore better adapted to dry climatic conditions. *Acacia* species have globular or cylindrical flower heads made up of many individual flowers. These flowers have only tiny petals, the prominent parts being the stamens that give them their 'fluffy' appearance.

Note: We have divided this large and confusing genus into groups that show similar-looking species together. The species are not necessarily taxonomically close.

Large shrubby wattles

2 ***Acacia cyclops***, Coastal Wattle, Red-eyed Wattle **05595002**

In Greek mythology the Cyclops were a race of one-eyed giants. The name alludes to the story of Odysseus who gouged out the eye of the king of the giants with a burning stake; the plant's seed is black, surrounded by an orange-red, circular aril or stalk, reminiscent of a large eye. This aril is attractive to birds that eat and disperse the seeds. *A. cyclops* is a tall, dense shrub to about 6 m found right around the coast from Shark Bay into South Australia. It grows vigorously on dunes and limestone sands and is an excellent dune stabiliser.

Flowers: Sept.–Mar.

3 ***Acacia cochlearis***, Rigid Wattle **05623928**

L. *cochlear* = spoon; refers to the shape of the sepals

The common name for this plant is the more descriptive as the pale green phyllodes of this erect, bushy shrub stick out stiffly at right-angles. It can be up to 3 m tall. It is another coastal species, growing on limestone or granite from Lancelin right around the south coast to the Nullarbor. It is one of the earlier flowering wattles.

Flowers: Aug.–Oct.

4 ***Acacia myrtifolia*** **05145643**

L. *myrtus* = myrtle + *folium* = leaf

A large bushy shrub to about 3 m with long racemes of creamy-yellow flowers. This is one of our most common wattles, found on sandy or gravelly soils in Jarrah forest and woodlands and conspicuously along roadsides. It is noticeable because of its early flowering time. It occurs from Perth to Israelite Bay.

Flowers: July–Nov.

5 ***Acacia rostellifera***, Summer-scented Wattle **05624177**

L. *rostellum* = small beak or snout + *fero* = to bear; the linear phyllodes are curved at the tips

It is sometimes tricky to distinguish between our various large wattles. The curved ends to the long, thin phyllodes of this one are the clue. This shrub can be up to 6 m tall, but more often is about 3 or 4 m, growing in dense thickets on sand over limestone. It spreads rapidly by means of suckers, making it another good dune stabiliser, and is very common around the Gracetown townsite. It extends around the coast from Shark Bay to Esperance.

Flowers: Oct.–Dec.

6 ***Acacia saligna***, Orange Wattle **05750008**

L. *salignus* = resembling willow wood

Another common tall shrub in the area, this wattle grows up to 6 m and is found in a variety of soils and habitats. The flowers are a deep orange-yellow and the phyllodes are dark and broad. They often show signs of heavy insect attack as do many other wattles and pea-flowers. Their ability to thrive due to their own home-made fertiliser also makes them attractive to leaf-eating insects. Widespread throughout the south-west.

Flowers: Aug.–Oct.

Note: It is interesting that *A. cyclops* and *A. saligna*, both introduced to South Africa as dune-stabilisers, have become rampant weeds in that country because they have no pests or diseases to keep them in check. This is 'tit for tat' with some of our worst weed offenders having travelled the other way.

P.Negus

Smaller wattles

1 ***Acacia alata*** var. ***alata***, Winged Wattle **05750024**

L. *alata* = winged (stems)

Though sometimes found in gravelly country this species prefers moist habitats. It is often present in the Karri forest where it is damp and shady and it likes growing along creek-lines where it grows tall and lanky to about 1.5 m with support from other plants. Like most wattles it grows prolifically after fire, but soon ages to a straggly, untidy appearance. It grows all down the west coast and adjacent inland areas from Kalbarri to Albany.

Flowers: July–Oct.

2 ***Acacia stenoptera***, Narrow-winged Wattle **05611830**

G. *stenos* = narrow + *pteron* + wing

A low, straggling shrub with rigid stems less than 1 m tall. It is another autumn flowerer and is found on sandy, lateritic soils in Jarrah forest, or on damp, sandy flats. I have not yet found its reddish, quadrangular-shaped pods. Common from Geraldton to Jerramungup.

Flowers: April–June

3 ***Acacia obovata***, Wavy-leaved Wattle **05611857**

L. *ob* = inverted + *ovatus* = egg-shaped; refers to the phyllodes

The wavy, grey-green phyllodes of this little wattle are its distinguishing feature. It is a small branching shrub usually less than 50 cm high and is not very noticeable in the bush until it flowers in autumn. The flowers are cream to pale yellow. It grows on gravelly, lateritic soils in Jarrah forest and woodland between Jurien Bay and Augusta.

Flowers: Mar.–June

4 ***Acacia urophylla***, Net-leaved Wattle **05460891**

G. *ouro* = tail + *phyllus* = leaf

This is an erect, open shrub to 2 m or more with characteristic, prominently-veined phyllodes and prolific flowers along the stems. It is found on loamy or lateritic soils in Jarrah and Karri forest and is common along the walk-trail from Rotary Park in Margaret River to the weir. It occurs from well north of Perth down to Albany.

Flowers: Sept.–Nov.

5 ***Acacia scalpelliformis*** (leaves only illustrated below)

L. *scalpellum* = small surgical knife (scalpel) + *forma* = shape; refers to the phyllodes

This is very similar to *A. urophylla,* but has smaller, more pointed phyllodes with an acute angle at the top. It grows in the same type of habitat and the two species are often found together, although *A. scalpelliformis* appears to have a more restricted distribution between Margaret River and Denmark. It has been suggested that it may, in fact, be a hybrid between *A. urophylla* and *A. divergens.*

Flowers: Oct.–Nov.

6 ***Acacia extensa***, Wiry Wattle **05750016**

L. *extensus* = stretched out

The phyllodes of this species are very long and terete (rounded), resembling branchlets more than leaves. It is a slender, spreading shrub to about 2 m, but often broader than it is high. It grows on lateritic or sandy soils in Jarrah forest, often in wetter areas, and is found from north of Perth to Albany.

Flowers: Aug.–Oct.

5 ***Acacia scalpelliformis***

P.Negus

Scratchy wattles with sharp, angular phyllodes!

1 *Acacia divergens* **05316200**

L. *dis* = asunder, away from + *vergo* = to turn; spreading outwards

A diffuse, open, often weeping plant reaching up to 2 m tall with pungent (sharp) phyllodes. The flowers have small, rounded calyx lobes. It grows on sandy clay soils in a variety of habitats, but usually in damp situations on swamp edges or along watercourses. It is common in Jarrah forest between Perth and Bremer Bay.

Flowers: Oct.–Nov.

2 *Acacia mooreana* △ **05145708**

After Sir Newton James Moore (1870-1936), premier of WA 1906-10.

An erect, slender shrub up to 1 m high. This also has sharp, angular phyllodes and can be confused with *A. divergens*, especially as it tends to grow in the same locations. It is a smaller plant, upright rather than weeping, with a slightly earlier flowering time. If you look at the individual flowers with a lens, particularly the buds, you will see they are quite different from those of *A. divergens*, being much longer with pointed calyx lobes. The flower stalks, or peduncles, are also shorter. It is found between the Capes coast and Albany.

Flowers: Sept.–Oct.

3 *Acacia littorea*, Coastal Wattle, Shark's Tooth Wattle **05460980**

L. *litoreus* = pertaining to the sea-shore

This is the common wattle on the calcareous sandy soils of the Leeuwin-Naturaliste Ridge and its prolific flowers dominate large tracts of the coastal heath in spring. It is a dense shrub varying in height from a wind-pruned few centimetres in exposed locations to 2 m or more in sheltered swales. It is also very variable in form, from thick and chunky near the sea, to slender and weeping in forest shade. It occurs from Busselton around the coast to Bremer Bay, but is also found on Rottnest Island. Around Perth there is a very similar, but less prickly species, *A. truncata*.

Flowers: Aug.–Nov.

4 *Acacia uliginosa* **05474329**

L. *uliginis* = moisture + *-osus* = abounding in; refers to its habitat

A slender, sprawling small shrub usually less than 50 cm tall. It is unobtrusive except when flowering when its mass of pale flowers along the stems is quite striking. Its short, but sharp-pointed phyllodes make this a very prickly little plant. It grows on sandy or clayey soils in damp places and occurs along the coastal plain from Mandurah to Albany.

Flowers: Aug.–Oct.

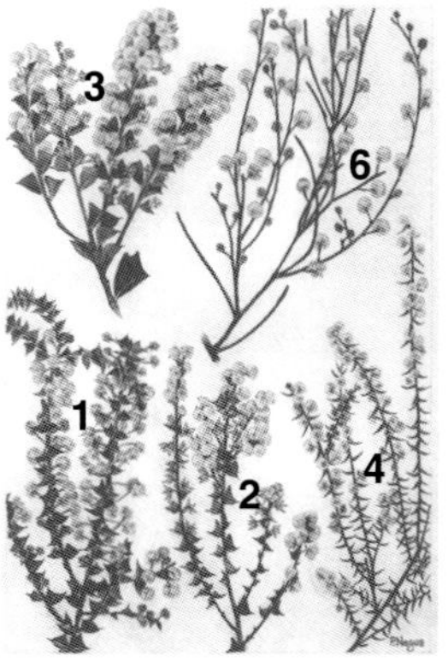

P.Negus

Wattles with real leaves

1 ***Acacia gilbertii*** **05335868**

After John Gilbert (1812-45), English naturalist and explorer, who visited WA in 1839-40 and 1842-43 to collect birds for John Gould.

This is a pretty little wattle not much more than 1 m high with delicate leaves, reddish stems and twisting red-brown pods. It flowers late, in summer, and has creamy-white instead of the more usual yellow flowers. Found on gravelly soils in Jarrah forest from Bindoon to Albany.

Flowers: Dec.–Feb.

2 ***Acacia pulchella*** var. ***pulchella***, Prickly Moses **05474671**

L. *pulchellus* = small and beautiful

This well-known small shrub generally less than 1 m high is found on both limestone and lateritic soils in Jarrah forest and coastal woodlands. It makes a showy display in early spring, especially after fire, when its spiny branches form impenetrable thickets in the bush. Widespread throughout the south-west.

Flowers: July–Sept.

3 ***Acacia browniana*** var. ***browniana*** **05316413**

After Robert Brown (1773-1858), naturalist on Flinders' expedition and the most distinguished early collector of WA flora.

There are several variants of this species and it is also similar to other, closely related species that grow further inland. This variety is an erect and spreading shrub to 2 m found in Jarrah forest on sandy or lateritic soils, in low-lying areas, slopes and granite outcrops. Where present it is often a dominant under-shrub.

Flowers: Aug.–Oct.

4 ***Acacia varia*** var. ***varia*** **05456916**

L. *varius* = variable

This is an early flowering species of inland areas that grows in profusion on lateritic soils in Jarrah forest from Perth to Albany. It is found in the north-east of our area at the western limit of its range. It is a small, erect shrub only reaching about 50 cm tall.

Flowers: July–Sept.

5 ***Acacia subracemosa***, Western Karri Wattle △ **05772176**

L. *sub-* = somewhat + *racemosus* = in a raceme (of flowers)

This tall shrub or small tree growing to 5 m is a local inhabitant of the Karri forest of the Leeuwin-Naturaliste Ridge. It favours sandy, limestone soils. It is a prominent understorey species here, but is restricted to this small outlier of Karri. A different species, *A. pentadenia*, takes its place throughout the main belt of Karri forest in the Pemberton–Walpole district.

Flowers: Oct.–Nov.

6 ***Acacia*** **aff.** ***lateriticola*** (not illustrated) **05456967**

L. *later* = a brick + *-ita* (denotes a connection), hence laterite, + *-cola* = inhabitant

This small shrub to about 1 m is found in the north-east of our area, growing with *A. varia* var. *varia* on laterite soils. It has pinnate leaves and looks somewhat similar, but has globular instead of cylindrical flower heads, and flowers a little later. It may be a variant of *A. lateriticola*, but has hairy instead of smooth peduncles (flower head stalks).

Flowers: Aug. – Oct.

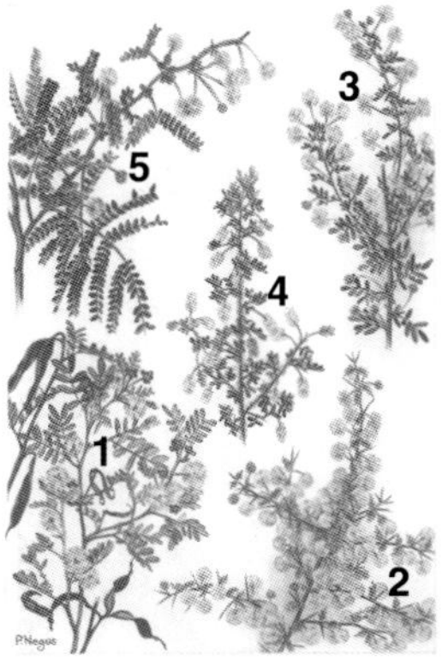

P.Negus

Papilionaceae, Pea Flowers

Like the wattles, this is a family of pod-bearing shrubs and herbs with nitrogen-fixing bacteria in their root-nodules. It is one of the biggest plant families in the world and in the south-west of WA there are about 40 native genera as well as several introduced ones (see weeds). Many of the pea-flowers are brilliantly coloured and range throughout the rainbow spectrum from red to purple. Some are small and solitary while others are abundant and showy when in flower. They may be shrubs, herbs or climbers but all have the characteristic 'pea' flower shape.

Mirbelia

After Charles Francois Brisseau de Mirbel (1776-1854), botanist and professor of culture at the museum of Paris.

1 ***Mirbelia dilatata***, Holly-leaved Mirbelia **05335841**

L. *dilatatus* = enlarged, widened

The leaves of this plant are widened at the ends into three to seven very prickly lobes. It is, however, very attractive when flowering, especially as it tends to grow in dense stands. It is a spreading shrub up to 3 m tall and likes sandy and gravelly soils in Jarrah forest. It is a coloniser and grows well on disturbed sites such as abandoned gravel pits and roadsides. Widespread throughout the south-west from Perth to Esperance.

Flowers: Oct.–Jan.

Callistachys

G. = ear of corn, referring to the flower head

2 ***Callistachys lanceolata***, Native Willow **05771889**

L. *lanceolatus* = shaped like a lance-head, narrow with curved edges tapering to a point; refers to the leaves

This species was classified within the genus *Oxylobium*, but has had its old generic name, *Callistachys*, reinstated. It is a tall shrub or small tree to 8 m tall often with weeping branches and leaves in whorls. It grows on sandy soils in damp situations along watercourses and swamps, appearing rather like a European willow, although they are completely unrelated. It is a common tree in our area and is easy to see, for instance, along Wallcliffe Road towards Prevelly, just west of Caves Road. It is found between Perth and Cape Arid.

Flowers: Sept.–Jan.

Oxylobium

G. *oxys* = sharp + *lobos* = pod

3 ***Oxylobium lineare***, Narrow-leaved Oxylobium, River Pea **05577322**

L. *linearis* = linear

This shrub looks rather similar to the previous one and shares similar damp habitats, but is smaller with a loose raceme of red and yellow rather than orange-yellow flowers. Its narrow leaves are alternate and its pods are smaller. It is found from the south coast to Perth with isolated occurrences further north as far as Kalbarri. This species is soon to be transferred to the genus *Gastrolobium*.

Flowers: Oct.–Jan.

Gastrolobium, Poison Peas

G. *gastros* = paunch, belly + *lobos* = pod

4 ***Gastrolobium bilobum***, Heartleaf Poison **05772141**

L. *bi-* = two + *lobus* = lobe (leaves)

Plants of the genus *Gastrolobium* contain the chemical fluoroacetate from which the poison 1080, so successful in controlling feral animals in Western Australia, is derived. Native mammals in WA are to a large degree immune to its effects, whereas ferals such as foxes, rabbits and domestic dogs have no resistance. Eastern States native mammals are unfortunately also susceptible. We have only the one species of *Gastrolobium* here and it is common on granite outcrops. It is as toxic as any so take care not to get any in your mouth. Heartleaf is widespread between Perth and Albany and extends along the south coast to Esperance.

Flowers: Sept.–Nov.

Isotropis

G. *isos* = equal + *tropis* = keel; the keel petal is about as long as the wings.

5 ***Isotropis cuneifolia*** subsp. ***cuneifolia***, Granny Bonnets **05624150**

L. *cuneus* = wedge + *folium* = leaf

A small, creeping or ascending herb, or slightly woody shrub only a few centimetres high. It grows on sandy or lateritic soils over granite or limestone in open woodland or heath. You need to closely examine the back of the petals of the flowers to see the exquisite veins that are the most striking feature of this lovely little plant. It is found between Geraldton and Esperance and through the wheatbelt. It may be toxic to stock.

Flowers: Aug.–Nov.

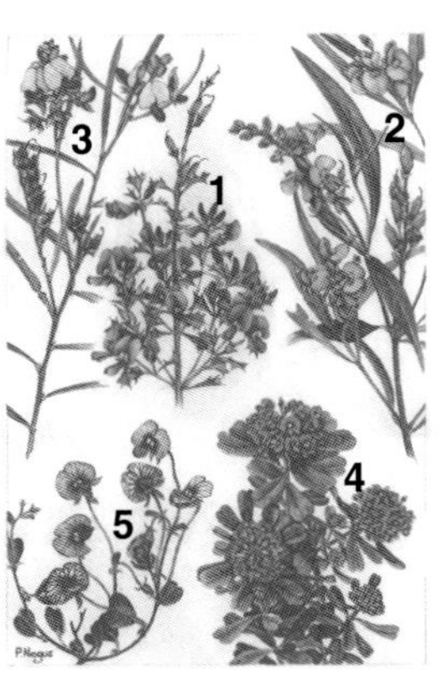

P.Negus

Chorizema, Flame Peas

G. *chorizo* = to divide + *nema* = thread; referring to the stamens that are separate, not united in a sheath. Note: The origin of names often becomes clouded with history and there is another, perhaps more captivating, version of how *Chorizema* was given its name by the botanist, Jacques Julien Houton de le Billardiere. It is said that as his expedition travelled down the WA coast they became very short of water. When they finally found a spring he danced for joy and named the bright little pea-flower that he found growing beside it *Chorizema*, meaning: G *choros* = to dance + *zema* = a drink!

The flame-peas are generally small shrubs or twiners, but they are all brilliantly-coloured in reds, pinks, oranges and yellows. Each one illustrated here has a different hue.

1 ***Chorizema aciculare*** subsp. ***aciculare***, Needle-leaved Chorizema **05625068**

L. *acicula* = a small pin

This is a lovely small spreading shrub to about 50 cm that is found on lateritic and sandy soils. I have so far only found a small patch in our area on granite outcrops near Augusta, but it is widespread further inland and is found from Geraldton to Cape Arid.

Flowers: Sept.–Oct.

2 ***Chorizema diversifolium*** **05577403**

L. *diversus* = different + *folium* = leaf

The leaves of this slender trailing or climbing shrub vary considerably in width. It is sometimes confused with Coral Vine, *Kennedia coccinea,* that has similarly-coloured flowers. It is a more delicate plant, however, with fewer flowers in long, open racemes instead of the dense heads of Coral Vine. The flowers are also more orange than red. Common in Jarrah and Karri forest on a variety of soils, it occurs through the lower south-west to just east of Albany.

Flowers: Sept.–Dec.

3 ***Chorizema nanum*** **05750032**

L. *nanus* = dwarf

This is one of a group of closely-related small shrubs, all with broad spiny leaves that can be difficult to differentiate. *C. cordatum* is a larger plant, and *C. ilicifolium,* which may also be in the area, has more pointed leaves. *C. nanum* prefers sandy or clay loam soils in Jarrah forest or heath and is found only in the far south-west corner, whereas *C. cordatum* occurs in the Darling Range further north.

Flowers: Sept.–Nov.

4 ***Chorizema reticulatum***, Showy Flame Pea △ **05460948**

L. *reticulatus* = net-veined (leaves)

Though not large, this erect, wiry shrub to about 50 cm has delicate pale pink-orange flower spikes that are conspicuous amongst the low understorey plants of the Jarrah forest. It grows on sandy lateritic soils between Busselton and Albany.

Flowers: Aug.–Oct.

5 ***Chorizema rhombeum*** **05316227**

G. *rhombeus* = diamond-shaped (leaves)

Low, prostrate or ascending shrub less than 50 cm high growing on sandy, clay or gravel soils in Jarrah forest. The flowers are usually sparse, but stand out like little red jewels in the grey-green bush. It is found from Perth to Albany.

Flowers: Sept.–Oct.

1 ***Chorizema aciculare*** subsp. ***aciculare***, Needle-leaved Chorizema

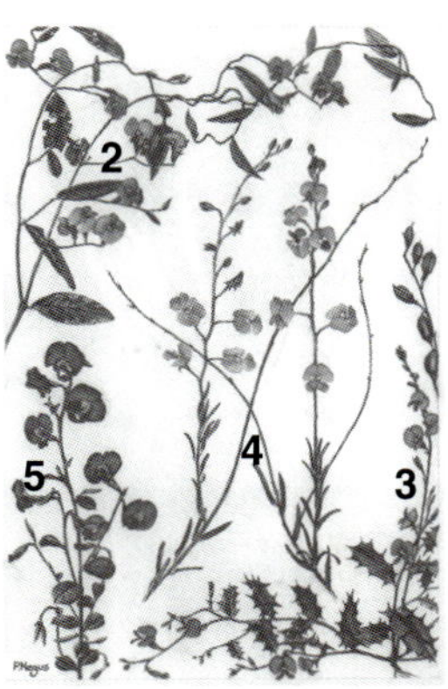

P.Negus

Gompholobium, Wedge Peas

G. *gomphos* = club + *lobos* = pod

This is a genus of small to medium shrubs and twiners that come in a wide variety of bright colours.

1 *Gompholobium capitatum*, Yellow Pea **05515424**

L. *capitatus* = forming a head

A low-growing spreading shrub to 1 m high with a mass of yellow flower heads. It grows in sandy heathland and open woodland between Pinjarra and Albany.

Flowers: Oct.–Dec.

2 *Gompholobium confertum* **05577330**

L. *confertus* = crowded

This stiffly erect shrub that reaches about 1 m high has crowded needle-like leaves and dense purple flower heads. It is so conspicuous that it looks more like a weed that escaped from a garden somewhere, but it does belong to our bushland. It grows on sandy and lateritic soils and is found from Eneabba to Esperance.

Flowers: Oct.- Nov.

3 *Gompholobium knightianum* **05460921**

After Joseph Knight (?1781-1855), gardener at King's Road nursery, Chelsea in London; this species was described from a specimen grown at the nursery.

A diminutive, slender plant less than 50 cm high, but with large, eye-catching mauve flowers and pinnate leaves. It is common on gravelly soils throughout the Jarrah forest. A closely-related species, *G. venustum,* with lovely blue and pink flowers, grows in a similar habitat further inland.

Flowers: Oct.–Nov.

4 *Gompholobium ovatum* **05750040**

L. *ovatus* = egg-shaped (leaves)

The large flowers of this *Gompholobium* come in many different colours; yellow, orange, red, brown, purple or pink! Here they are generally a beautifully rich orange. Another low-growing species of the Jarrah forest floor found between Bunbury and Albany.

Flowers: Oct.–Nov.

5 *Gompholobium polymorphum* **05335809**

G. *polys* = many + *morphe* = form

This is an interesting plant with, as its name implies, various different forms. It can be a bushy little shrub or an elongated climber, can have yellow or orange-red flowers – I have only found yellow in this area– and has varying leaf shapes; how confusing! It is generally only about 50 cm high, except in its climbing phase and is found on lateritic soils in Jarrah forest from Moora to Albany and along the south coast to Esperance.

Flowers: Oct.–Dec.

6 *Gompholobium preissii* **05624983**

After Johann August Ludwig Preiss (1811-83) who collected upwards of 2,700 species between 1839 and 1841.

This is a small shrub only about 30 cm high with orange-yellow flowers and is hard to see amongst the leaf litter on the Jarrah forest floor. It is usually found on lateritic, gravelly soil and ranges from Eneabba to Albany.

Flowers: Sept.–Nov.

7 *Gompholobium tomentosum*, Hairy Yellow Pea **05515300**

L. *tomentum* = woolly padding, densely matted hairs

An erect shrub usually less than 1 m high. It has some similarities with *G. capitatum* with its bright yellow flowers and bushy habit and it shares similar, sandy heathland habitat. However, its flowers are paler yellow and there are fewer per head. It is also a taller, bushier shrub. Widespread from Shark Bay to Esperance.

Flowers: Oct.–Dec.

6 *Gompholobium preissii*

P.Negus

Jacksonia

After George Jackson (?1780-1811), Scottish botanist

Jacksonia is a large genus of well over 40 species and is endemic to WA. They are shrubs or small trees, often leafless, but sometimes very floristic. They mostly have yellow-orange flowers and grey-green foliage.

1 *Jacksonia horrida* **05577268**

L. *horridus* = shaggy, prickly, very thorny

The meaning aptly describes this species! It is a tall shrub growing to about 2.5 m with straggly, weeping branches and a lot of prickles, though it can be quite an elegant plant when young and surprisingly soft in coastal locations. Its branches become narrower, spinier and less flattened away from the coast. It flowers prolifically in late spring and summer, but a few flowers are usually present all year round. It mostly, but not exclusively, grows on sandy soils over coastal limestone between Mandurah and Albany.

Flowers: all year

Viminaria

L. *viminis* = withe, osier (willow used in basketwork); refers to the rush-like stems

2 *Viminaria juncea*, Swishbush, Golden Spray **05335833**

L. *junceus* = rush-like

There is only one species in this genus. *Viminaria* is a widespread and common plant of forest and woodland between Kalbarri and Esperance. It grows well on clay soils in wet areas and disturbed habitats. It is generally a tall shrub up to 4 m high with slender, weeping branches, though in windswept coastal locations, such as out on the cliffs north of Gracetown, it can look quite different, growing squat and thick stemmed. The long sprays of yellow flowers, though, are distinctive.

Flowers: Oct.–Jan.

Sphaerolobium, Globe Peas

G. *sphaira* = ball, sphere + *lobos* = pod

This is a small genus of less than 20 species most of which are endemic to the south-west of WA. They are all slender, almost leafless shrubs with elongated racemes of flowers, developing into rows of small, globular pods.

3 *Sphaerolobium drummondii* **05515246**

After James Drummond (1784-1863), very successful collector and first Government Horticulturist for WA, 1829-34.

This rigid, leafless plant is only about 70 cm high, but has spectacular orange flower spikes. It favours sandy clay and loam soils in dry heathland or on granite outcrops and is found around the coastal plain between Kalbarri and Esperance.

Flowers: Sept.–Nov.

4 *Sphaerolobium fornicatum* **05577233**

L. *fornicatus* = arched, vaulted; refers to the very curved keel petal of the flower

The flower spike of this plant looks similar to the above species, but it is much taller and more slender. It is not a climber, but depends on the surrounding vegetation for support and can reach 1.5 m amongst the dense swamp vegetation where it grows. Found in swamps around the coast from Mandurah to Albany.

Flowers: Oct.–Dec.

5 *Sphaerolobium medium* **05316243**

L. *medius* = common, middle

This is the *Sphaerolobium* species that you are most likely to see. It is a common small shrub to 50 cm high of open Jarrah woodland growing on sandy or lateritic soils. It has a clumped habit and grows in extensive stands with conspicuous yellow or orange flower spikes. It is especially prevalent in areas recently burnt. Widespread between Jurien Bay and Albany through the Jarrah forest and coastal plain.

Flowers: Sept.–Nov.

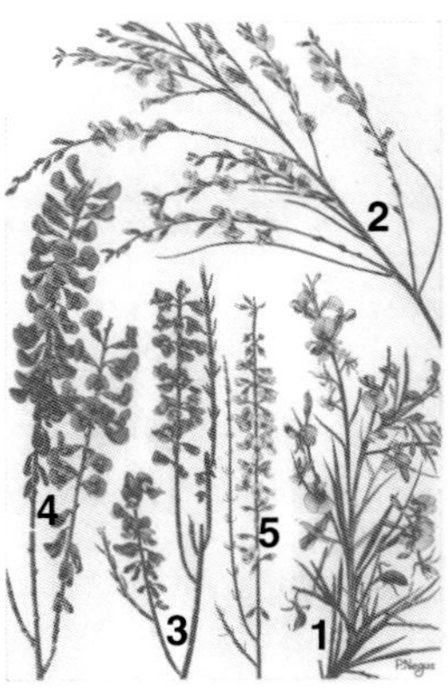

P. Negus

Daviesia, Bitter Peas, Rattle Peas

After Hugh Davies (1739-1821), Welsh botanist

Daviesia is a distinctive genus, its members having large numbers of small, crowded flowers that are generally colourful combinations of red, orange and yellow. The pods are triangular and inflated when ripe with the seeds often rattling inside before they split open. In *Daviesia* the leaves are replaced by phyllodes (flattened stems) that in many species are pungent (sharp pointed).

1 *Daviesia cordata*, Bookleaf **05474353**

L. *cordis* = heart; heart-shaped (phyllodes)

This shrub is one of the larger species of *Daviesia*, growing up to 1.5 m. It has showy terminal flower clusters with surrounding bracts that enlarge after flowering to enclose the developing pods, giving the plant its common name. It grows on sandy lateritic or granite soils, usually in drier parts of the Jarrah forest and is found between Bindoon and Bremer Bay.

Flowers: Sept.–Nov.

2 *Daviesia decurrens*, Prickly Bitter-pea **05145635**

L. *decurrens* = running down; the lower part of the phyllode runs down the stem

This plant shares its common name with one or two others (see below). This is one good reason for getting to know the unique scientific name for each individual species. *D. decurrens* is a fairly easy plant to recognise with its clumped habit and its mass of flame-coloured flowers in late winter and early spring. Its lovely green-red pods are almost as decorative as the flowers. It is less than 1 m high and prefers lateritic soils in Jarrah forest, but occurs from Geraldton to east of Albany and inland as far as Kalgoorlie.

Flowers: July–Sept.

3 *Daviesia horrida*, Prickly Bitter-pea **05460808**

L. *horridus* = shaggy, prickly

This is, indeed, a dense prickly plant with rigid spines growing to about 1.5 m on laterite or granite outcrops and rocky hillsides. Its blue-green stems contrast with the mass of brilliant orange-red flowers. As well as its spiny, leafless branches this plant has occasional broad, flat phyllodes. It has a patchy distribution between Perth and Albany.

Flowers: Aug.–Nov.

4 *Daviesia inflata* **05474345**

L. *inflata* = swollen; refers to the very rounded pods when ripe

Another medium shrub of the jarrah forest growing to about 1.5 m with long terete (rounded) stems and short, narrow, pungent phyllodes. Its many stems bear masses of bright orange-red flowers. It is a more open bush than *D. decurrens* and is later flowering and taller. It is found through the lower south-west from Bunbury to Esperance.

Flowers: Sept.–Nov.

5 *Daviesia preissii* **05456959**

After Johann August Ludwig Preiss (1811-83) who, with Drummond, was one of the most prolific collectors of WA plants

This is a small dense shrub generally less than 50 cm. Once again it has short, rigid, spiny phyllodes that are slightly flattened. The yellow and red flowers, nestled in the axils, are sparser than in many *Daviesia* species and appear later than most, in summer. It grows on sand, clay or gravel hill slopes in Jarrah forest from Perth to Albany, but mostly further inland.

Flowers: Dec.–Feb.

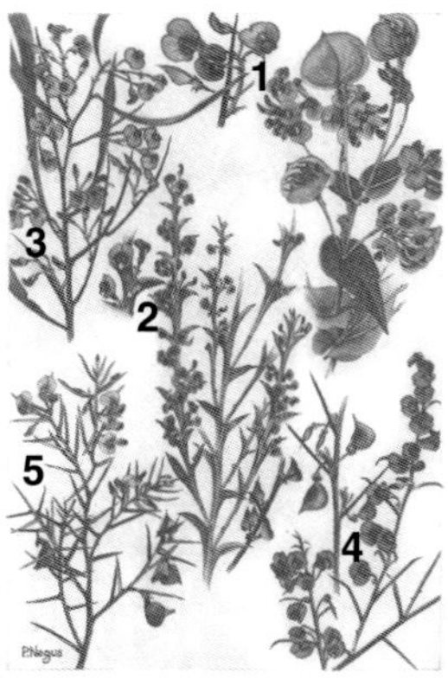

P.Negus

'Eggs and bacon' peas

This common name refers to any number of the orange-yellow-brown pea-flowers occurring in the south-west of WA, which can cause great confusion. The various plants on this page, as well as species of the genus *Bossiaea*, come into this category. The illustrations and descriptions here should make their differences a little clearer.

Pultenaea

After Richard Pulteney (1730-1801), English botanist

Medium shrubs with yellow-orange flowers, usually in heads.

1 *Pultenaea drummondii* **05316251**

After James Drummond (1784-1863), first Government Horticulturalist, 1829-34 and a great collector of WA plants

This is a slender, open shrub to just over 1 m that grows in extensive stands in Jarrah forest on sand, clay or laterite soils on flats or swamp edges. Although common it has a fairly localised distribution between Busselton, Nannup and Augusta.

Flowers: Sept.–Nov.

2 *Pultenaea reticulata* **05461057**

L. *reticulatus* = net-veined (leaves)

Larger than the previous species, *P. reticulata* is a stout, bushy shrub with pungent leaves, growing up to 2 m tall. It is a common shrub of large areas of the sandplain country behind the limestone ridge in the Prevelly and Calgardup areas, but it is also found in open Jarrah woodland further east. It is especially prevalent in the lower south-west from Busselton to Albany, but occurs as far north as Jurien Bay.

Flowers: Sept.–Nov.

Eutaxia

G. *eutaxia* = good order, discipline; probably refers to the regularity of the leaves.

Small to medium shrubs with neat, opposite leaves and yellow-brown flowers.

3 *Eutaxia parvifolia* (not illustrated) **05772109**

L. *parvus* = small + *folium* = leaf

This plant grows out on the granite cliffs north of Gracetown where it is wind-pruned to a low ground-cover. In sheltered places it may grow to 50 cm or more. It is found from Perth to Esperance, but with a generally more inland distribution.

Flowers: Aug.–Oct.

4 *Eutaxia obovata* **05772001**

L. *ob* = inverted + *ovatus* = egg-shaped

This is also a plant of granite outcrops, but is a much larger more open shrub up to 2 m tall with larger leaves and flowers and a later flowering period. Mostly found between the Capes and Esperance.

Flowers: Sept.–Dec.

5 *Eutaxia virgata* **05577608**

L. *virgatus* = twiggy; slender and stiff

'Twiggy' describes this species well, at least in the form it takes locally. It varies somewhat in appearance over its range, which is between Perth and Albany. It is a straggling, slender shrub up to 1 m high with tiny flowers hugging the stems. It grows on sandy and clayey soils in swamps, carpeting the ground around the swamp edges as they dry out in spring.

Flowers: Nov.–Feb.

Dillwynia

After Lewis Weston Dillwyn (1778-1855), English botanist

Closely related to Eutaxia, but with alternate or scattered leaves.

6 *Dillwynia* sp. *A Perth Flora (R. Coveny 8036)* **05460778**

This species has still not been formally named although it is common in the south-west. It was for many years mistaken for a very similar Eastern States species, *D. cinerascens*, but is now recognised as being distinct. It is a dense, prickly shrub up to 1.5 m with spiny branchlets growing on granite outcrops, laterite hill slopes or winter-wet flats. Its masses of orange-yellow flowers make it a prominent indicator of granite outcrops throughout our area when it is in bloom. Found from north of Perth to Albany.

Flowers: Aug.–Oct.

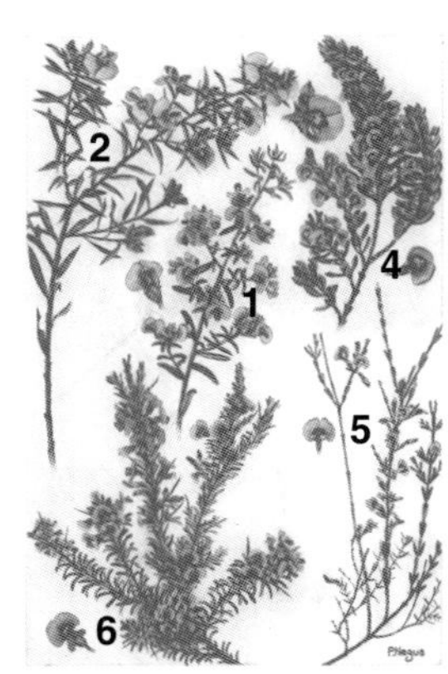

P.Negus

Hovea

After Anton Pantaleon Hove; Polish plant collector for Kew Gardens in West Africa, the Crimea and India.
Hovea is a genus of around 20 species, endemic to Australia, with just eight species found in the south-west. They are all purple.

1 *Hovea chorizemifolia*, Holly-leaved Hovea **05145503**

From the genus *Chorizema* + L. *folium* = leaf

This is an example of the roundabout way plants are sometimes named. Several, though by no means all, *Chorizema* species have prickly, holly-like leaves. The large, brilliant flowers of this species brighten the Jarrah forest in the middle of winter. The plant itself is quite small, around 50 cm and is common on sand or gravelly soils. Common throughout the jarrah forests from Perth to Albany.

Flowers: July–Sept.

2 *Hovea elliptica*, Tree Hovea **05316332**

G. *ellipticos* = like an ellipse (with pointed end)

A tall shrub to 3 m with broad, elliptical leaves. The flowers look similar to *Hardenbergia* (see below) and the two are often confused, but of course *Hardenbergia* is a climber. When in full bloom, stands of *H. elliptica* make one of the south-west forests' most spectacular sights. It grows on a variety of soils in both Jarrah and Karri forest and is a prominent species of the wetter south-west between the Capes and Bremer Bay.

Flowers: Sept.–Nov.

3 *Hovea pungens*, Devils' Pins **05460824**

L. *pungens* = pricking; ending in a sharp, hard point

An erect, open shrub that may grow quite tall to 1.5 m though is often lower in our heathland. It usually has only a few stems that flower prolifically. It grows on a variety of soils over limestone or granite, in sandplain country or open woodland between Geraldton and Cape Arid.

Flowers: Sept.–Oct.

4 *Hovea trisperma*, Common Hovea **05320704**

G. *treis* = three + *sperma* = seed; refers to the three-celled ovary

Usually a straggly little shrub less than 50 cm with less prolific flowers than the other species and it is certainly less noticeable. This is another widespread species growing on a variety of soils in the Jarrah forest or coastal woodland and on heathland east to Esperance.

Flowers: Sept.–Nov.

Hardenbergia

After Franziska, Countess von Hardenberg, an active patron of botany and sister of Baron von Hügel

5 *Hardenbergia comptoniana*, Native Wisteria **05832632**

After Mary, 1st Marchioness of Northampton – Compton was the family name – who introduced this plant into cultivation in about 1810

One of our most well-known and best-loved native plants. It has been in cultivation for a long period, which is not surprising, given its spectacular flowering. It produces wonderful displays in the bush sometimes combined with Clematis and Coral Vine, or draped over *Hibbertia* or Wattle. It is a vigorous climber and grows on sandy soils around the coastal plain, Jarrah and Karri forest from Jurien Bay to Albany.

Flowers: July–Oct.

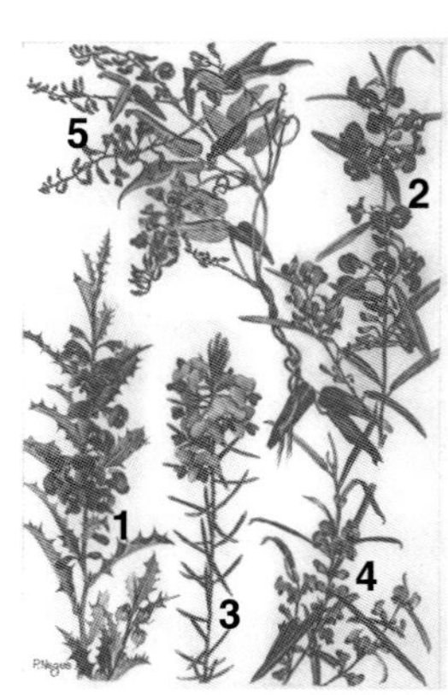

P.Negus

Bossiaea, Water-bushes, Brown Peas

After Joseph de Bossieu de la Martiniere (1754-88), medical officer and botanist on board the Astrolabe, La Perouse's voyage, 1785-88

Species of this genus range from very small to tall shrubs and differ considerably in their flowers and leaves. Most have masses of yellow-brown flowers, but some are less floriferous and a few are red.

1 ***Bossiaea aquifolium*** subsp. ***aquifolium***, Waterbush **05623960**

Pre-Linnaean name, Aquifolium, meaning holly; ie. holly-leaved

This species occurs in the north of the area in Jarrah woodland around Cowaramup. It is a shrub or tree that can reach 8 m in height though here is generally less than 2 m. It is a mass of yellow flowers in spring. This subspecies is found through the Jarrah forests while the subsp. *laidlawiana,* which does not occur here, is the common type found further south through the main belt of the Karri forest between Nannup and Albany.

Flowers: Sept.–Nov.

2 ***Bossiaea disticha*** △ **05461022**

G. *distichos* = in two rows (the leaves)

B. disticha is a familiar plant locally, being one of the most prominent understorey shrubs of the Boranup Karri forest. It is, however, very restricted in distribution, only occurring from Margaret River to Augusta and mainly west of Caves Road. It is a medium-sized spreading shrub to about 1.5 m and is found on sandy soils overlying limestone. It is found in coastal woodland as well as in the forest.

Flowers: Aug.–Oct.

3 ***Bossiaea linophylla*** **05316375**

G. *linon* = flax, thread + *phyllon* = leaf

Another prominent member of the Jarrah and Karri forest understorey this species is sometimes confused with the previous one, but has a more slender, weeping habit. The leaves are long, narrow and linear. It is also taller, up to 2.5 m. Again, it grows on sand over coastal limestone, but is also found on granite. It has a much wider distribution through the southern Jarrah forests.

Flowers: July–Nov.

4 ***Bossiaea ornata***, Broadleaved Brown Pea **05316316**

L. *ornatus* = handsome, showy

An erect, spreading shrub to about 1 m high growing on sandy and lateritic soils in Jarrah forest. This is another common species easily distinguished from the ones above by its larger yellow-brown flowers and broad, ovate leaves. It grows between Moora and Albany.

Flowers: Sept.–Nov.

5 ***Bossiaea praetermissa*** **05625157**

L. *praetermissa* = left out, neglected, omitted. Although it was first collected 200 years ago, this plant was named only recently, having been overlooked for many years.

Not all *Bossiaea* species are prominent bushes and this is a plant that is certainly easy to overlook! It is a prostrate or low-growing, thin, straggling shrub less than 1 m high with very small, solitary flowers along the stems. It is often leafless or nearly so. It grows on sand or clay soils in coastal heath and woodland, or on swamp margins and is found from the Capes eastwards to Esperance.

Flowers: Sept.–Nov.

6 ***Bossiaea rufa*** **05595061**

L. *rufus* = red

This species is similar in form to the previous with flattened stems and few or no leaves, but grows much taller, to 2 m with support from surrounding vegetation. It grows amongst dense vegetation on sandy, peaty soil in swamps and along stream banks and its bright red flowers are an attractive sight for anyone who ventures into their inhospitable environment. Found between Perth and Albany and along the south coast to Esperance.

Flowers: Nov.–Jan.

P.Negus

Templetonia

After John Templeton (1766-1825), Irish botanist

There are 12 species of *Templetonia*, eleven of which are in WA. We only have one, but it is one of the most spectacular.

1 *Templetonia retusa*, Cockies' Tongues **05145694**

L. *retusus* = blunt; the leaves having a rounded apex with a central notch

This is one of our most colourful winter flowers and it has a long flowering period. Flowers are generally bright red, but rarely may be white or yellow. This species is an indicator of limestone areas and grows in coastal heath and adjacent forest on sand over limestone all along the Leeuwin-Naturaliste Ridge. It is wind-pruned near the coast, but grows up to 3 or 4 m in sheltered locations. Found all around the coast from Shark Bay to the Great Australian Bight.

Flowers: April–Sept.

Kennedia

After John Kennedy (1759-1842), nurseryman at Hammersmith, London. Advised and supplied plants to the Empress Josephine

Another small genus of climbers and prostrate trailers with trifoliate leaves. Several are well-known in cultivation.

2 *Kennedia carinata* **05316340**

L. *carina* = keel, boat-shaped; the leaves have a raised central rib underneath

This is one of the more diminutive species in this genus. It is a prostrate ground-cover with dark green leaves and small but bright deep red or violet flowers fairly well-spaced along the stems. It creeps over the forest floor covering bare ground especially well after fires. Found from north of Perth to Albany.

Flowers: Sept.–Nov.

3 *Kennedia coccinea*, Coral Vine **05625262**

L. *coccineus* = scarlet

This vigorous climber festoons the bush with copious blooms wherever it grows, in forest, woodland or heath. It is especially noticeable after fires when its brilliant red flowers contrast starkly with blackened trunks and earth. It is common throughout the south-west between Perth and Esperance usually on sandy soils.

Flowers: Sept.–Nov.

4 *Kennedia macrophylla*, Augusta Kennedia

G. *macros* = large + *phyllon* = leaf

This plant is readily available from nurseries, but, sadly, is now known from only one or two locations in the wild. It is a superb species, a vigorous trailer or climber with large leaves and brick red flowers and grows on sandy, granitic soils.

Flowers: Sept.–Oct.

5 *Kennedia prostrata*, Scarlet Runner, Running Postman **05750059**

L. *prostratus* = lying along the ground

A prostrate or twining shrub with sparse but conspicuous flowers and large, crinkly leaves. It is common all around the west and south coasts from Shark Bay to Esperance, on sandy or gravelly soils in Jarrah forest or heathland.

Flowers: Aug.–Nov.

P.Negus

Geraniaceae

The genera of this family have delightfully connected names that liken the shapes of their fruits to the long bills of birds:
G. *Geranium* from *geranos* = a crane; *Pelargonium* from *pelargos* = a stork; and another one not represented here, *Erodium* from *erodios* = a heron.

Pelargonium

1 *Pelargonium capitatum, Rose Pelargonium **05750067**

capitatum L. = forming a head

This is an introduced species, but has long been an integral part of the coastal vegetation of the south-west. It is a shrubby perennial less than 1 m high with hairy leaves and stems that grows on sand-dunes and coastal limestone from north of Perth to Esperance. It is a native of South Africa.

Flowers: all year, but mostly Aug.–Nov.

2 *Pelargonium australe* subsp. ***drummondii***, Wild Geranium **05797721**

australe L = southern, Australian

An erect to sprawling perennial herb up to 40 cm high. It often grows on thin soils on granite outcrops and is found here on the cliffs above Cosy Corner. It is found along the south coast to Cape Arid. Specimens found here have considerably extended its known range westwards.

Flowers: Oct.–Dec.

3 *Pelargonium littorale* **05795141**

L. *littoralis* = belonging to the sea-shore

This is a soft-leaved, slender plant to around 50 cm that grows in the forest shade and has a rosette of small, pale flowers. It has a widespread distribution through the south-west.

Flowers: Oct.–Jan.

Geranium

4 *Geranium solanderi*, Native Geranium **05623952**

After Daniel Carl Solander (1733-82), favourite pupil of Linnaeus; accompanied Banks on Captain Cook's first voyage, 1768-71

Another small-flowered rather weedy looking perennial herb that is, however, native to the south-west corner where it is found between Perth and Esperance. It usually grows on sandy soils over either granite or coastal limestone.

Flowers: Aug.–Dec.

P. Negus

Zygophyllaceae

This is a worldwide family of trees and shrubs of some 240 species, of which we only have just over 30 in Australia. They are found mainly in arid and saline environments.

Nitraria

L. *nitrum* = saltpetre; refers to the salt-flats of Siberia where plants of this genus were first found.

1 *Nitraria billardierei*, Nitre Bush, Wild Grape **05456827**

After Jacques Julien Houton de le Billardiere (1755–1834), visited WA on board d'Entrecasteaux's voyage, 1791–94 and collected many species

A rigid, spreading shrub that can grow to 2 m with fleshy leaves and succulent fruit. It is found right around the south-west and southern coasts on coastal sand and limestone and is also found in the southern interior. There are small, isolated patches of it along our coast, for example at Gracetown.

Flowers: Sept.–Nov.

Euphorbiaceae

A large and diverse family distributed worldwide and especially in tropical regions. It includes such economically important plants as *Hevea brasiliensis*, source of most of the world's rubber and also manioc, or cassava. Flowers are mostly small, unisexual and often on separate plants. Many species exude a milky sap.

Beyeria

After A. de Beyer, early Dutch botanist who studied the ferns and mosses of Batavia

2 *Beyeria viscosa*, Pinkwood **05474485**

L. *viscum* = bird-lime + *-osus* = abounding in; viscid or sticky

A large, spreading shrub to 3 m with small inconspicuous flowers. Leaves are dark and shiny above, pale and waxy below (see *Diplolaena dampieri* p. 122). It has a patchy distribution in the area on coastal sands. It is found around the coast between Geraldton and Esperance.

Flowers: July–Sept.

Ricinocarpos

From *Ricinus communis* = the castor-oil plant (L. *ricinus* = a tick) + G. *carpos* = fruit; so: fruit is like that of the castor-oil plant, or a tick!

3 *Ricinocarpos glaucus*, Wedding Bush **05750113**

L. *glaucus* = bluish-grey or -green

This is an erect, open shrub about 1 m tall. Its beautiful white flowers have made it popular in cultivation. It is monoecious, having separate male and female flowers on the same plant. The female flowers are the larger. It grows on sandy soil over laterite or granite in open heath or woodland and is found scattered through the south-west from Kalbarri to Albany, especially prevalent in the south.

Flowers: Aug.–Nov.

Poranthera

G. *poros* = pore, passage + *anthera* = anther; the anthers open in pores

4 *Poranthera huegelii* **05515513**

After Carl Alexander Anselm, Baron von Hügel (1796-1870), Austrian traveller and naturalist

A small, erect, slender shrub about 50 cm high with flowers in loose panicles. It is common, but not particularly conspicuous in the Jarrah forests of the south-west between Perth and Albany.

Flowers: Oct.–Dec.

Phyllanthus

G. *phyllon* = leaf + *anthos* = flower; in the original species the flowering branches are leaf-like so the flowers appear to grow on leaves

5 *Phyllanthus calycinus*, False Boronia **05316448**

L. *calyx* + *-inus* = like; applied when the calyx is somehow unusual

This is a very familiar small shrub of our bushland, but its name is often unknown. It is usually monoecious with separate male and female flowers on the same bush, the female flowers being larger. It grows on sandy soils in Jarrah forest, woodland and heath. While it is green and healthy-looking through winter and spring, in summer it looks very stressed and often loses many leaves. It occurs right around the west and south coastline on limestone and also on granite soils of the Darling scarp.

Flowers: July–Dec.

♀
♂
♀ ♂
P.Negus

Rutaceae

Rutaceae is a large, worldwide family of interesting plants many of which have oil glands in their leaves. It includes the citrus fruits such as oranges and lemons. Some species have medicinal value and some, such as some of our *Boronias*, are cultivated for their oils for use in perfumes.

Boronia

After Francesco Borone (1769-94), assistant to several collectors and a most acute observer.

This genus is found only in Australia and many species are highly ornamental. Seven are illustrated here and there are probably more in this area. There are certainly others found not far to the east. Most species are pink or red.

1 *Boronia alata*, Winged Boronia **05750105**

L. *alatus* = winged (stems)

This is one of the characteristic heathland plants of our limestone coast, easily recognised by its strong distinctive smell when the leaves are rubbed. It has masses of beautiful flowers, usually pink, but occasionally white. It grows as a dense shrub on the coastal heath not much more than 1 m high, but in the shade of adjacent forest can be quite open and up to 3 m. It is found right around the coast from Perth to Esperance.

Flowers: Aug.–Nov.

2 *Boronia anceps* △ **05515238**

L. *anceps* = two-sided, two-edged, hence flattened (stems)

This species is restricted in its known distribution to a few sandy peat swamps near the coast in the far south-west corner. It is very similar to the more widely distributed *B. spathulata* found further inland, but has flattened stems. It is a small but striking shrub less than 1 m high with showy pink flowers.

Flowers: Sept.–Dec.

3 *Boronia crenulata* subsp. ***crenulata***, Aniseed Boronia **05461081**

L. *crenatus* = notched border + *-ul-* diminutive; divided into very fine teeth (leaf margins)

A spreading small shrub to about 1 m growing in diverse habitats on sand and clay soils in heath and forest. It has lovely deep pink flowers. This is a very variable plant with several sub-species. It is found between Perth and Ravensthorpe.

Flowers: Aug.–Nov.

4 *Boronia fastigiata* subsp. ***tenuior***, Bushy Boronia **05474388**

L. *fastigium* = gable; the stems, especially the pedicels (flower stalks) tend to grow erect and at an acute angle, maybe like an inverted gable

Another spreading shrub growing to about 1.5 m with attractive pale pink flowers and toothed leaves. It grows in seasonally damp sandy heathland or forest and along watercourses. The sub-species *tenuior* is found in the wetter south-west corner between the Capes and Walpole.

Flowers: Oct.–Dec.

5 *Boronia gracilipes*, Karri Boronia **05515378**

L. *gracilis* = slender + *pes* = foot; refers to the slender flower stalks

Many *Boronia* species are well-known plants; a fact that is demonstrated by the number that possess common names. This one occurs in the forest understorey of both Karri and Jarrah and, in fact, is more common in Jarrah forest in our area. It is a delicate open shrub up to 1.5 m tall growing on sand, loam or gravel soils. Found from the Capes across to Albany.

Flowers: Sept.–Dec.

6 *Boronia megastigma*, Scented Boronia **05456789**

G. *mega* = large + *stigma*

This is probably the best-known *Boronia* of all for its beautifully-perfumed yellow and brown flowers. Due to its popularity as a cut flower it is unfortunately not so common as it used to be and is now difficult to find in this area. It is an erect slender shrub to about 1 m and grows on peaty sand in winter-wet swamps of the lower south-west. It is usually hidden amongst dense vegetation, but is given away by its very distinctive scent.

Flowers: July–Sept.

7 *Boronia molloyae*, Tall Boronia **05515297**

After Georgiana Molloy (1805-43), one of the first settlers to Augusta and the Vasse and an avid and distinguished collector of plants, which she sent to horticulturalist, Captain James Mangles in England. She was the most important contributor to the early knowledge of flora in the Augusta-Margaret River area.

This is a beautiful slender shrub up to 3 m tall with gorgeous deep red flowers. It grows along streams and in swamps so is not easy to find. It grows mostly from Augusta east to Albany, but with outliers extending north as far as Perth.

Flowers: Sept.–Dec.

P.Negus

Diplolaena

G. *diploos* = double + laena (*chlaina* = cloak); refers to the double layer of bracts surrounding the flower head. This is made up of many individual flowers, with stamens much more prominent than the petals.

1 ***Diplolaena dampieri***, Southern Diplolaena **05750075**

After William Dampier (1652-1715), English navigator; collected the first Australian plants to reach European herbaria

A robust and common shrub of the limestone coast growing to about 2 m with red, orange or yellow flowers. The leaves are shiny and dark green above, pale and densely hairy below. This is an adaptation often seen in species growing on the coastal heath and helps to avoid desiccation from the strong, drying winds off the ocean. Southern Diplolaena is found along the coast from Perth to Augusta.

Flowers: Aug.–Nov.

Chorilaena

G. *chorizo* = to divide + *chlaina* = cloak; refers to the outer bracts (see *Diplolaena* above)

There is only the one species.

2 ***Chorilaena quercifolia***, Oak-leaved Chorilaena **05577381**

L. *quercus* = oak + *folium* = leaf

Oak-leaved Chorilaena is one of the main understorey trees of the Karri forest where it is found growing on sandy or loam soils, often on rocky hillsides of granite or limestone, up to 5 m high. But it is also found on the coast where it may be dense and wind-pruned to only 50 cm or less. The cream flowers, enclosed by green bracts, are difficult to see amongst the leaves. This is a plant of the wetter south-west, found between the Capes and Albany.

Flowers: Aug.–Dec.

Philotheca

G. *psilos* = naked, bald + *thece* = covering; the stamens are covered in long hairs, but are united at the base to form a glabrous (smooth) tube.

3 ***Philotheca spicata***, Pepper and Salt **05320712**

L. *spica* = point; refers to the tapering flower spike

Note: Used to be classed in the genus *Eriostemon*; all WA species are now included in *Philotheca*.

It is confusing when well-known species undergo a major name-change, but this reflects botanists' ever-increasing knowledge of the relationships between our plants. This one is a small spreading shrub about 50 cm high with prominent pink or mauve flower spikes. It is common in Jarrah forest, woodland and heath and is widespread in the south-west from Jurien Bay to Walpole.

Flowers: Aug.–Nov.

Rhadinothamnus

G. *rhadinos* = slender + *thamnos* = bush; refers to the columnar growth habit of the original species

4 ***Rhadinothamnus anceps***, Blister Bush **05771803**

G. *anceps* = two-sided, two-edged; the branches are angular

Here is another species that has recently changed its genus; it used to be called *Phebalium anceps*. It is a tall slender shrub up to 3 m growing on peaty sand along streams and in swamps. Take care when brushing through dense swamp vegetation as this plant can give the skin a nasty burn. It is found close to the coast in wetlands from north of Perth to Albany. Locally it is found along the damp area of the boardwalk above Ellensbrook Homestead. The flowers are attractive though and if you look closely, without touching, the fruits are interesting also.

Flowers: Oct.–Dec.

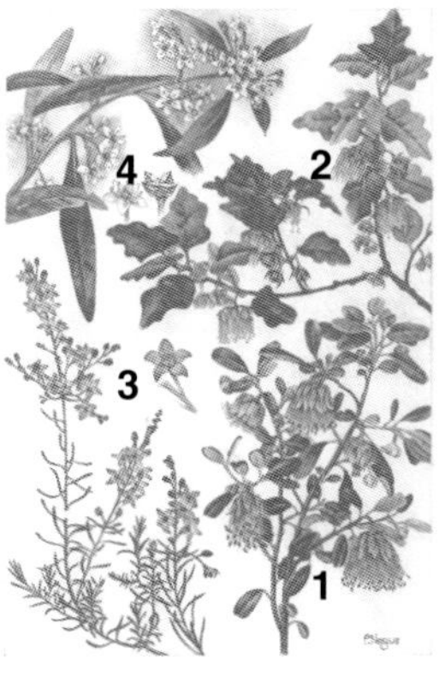

P.Negus

Tremandraceae

A small family, endemic to Australia with around 40 species in total. *Tetratheca* has the most species, of which 21 are found in WA.

Tremandra

G. *trema* = perforation + *andros* = man (bot. anther); refers to the difference between this and *Tetratheca*; in *Tremandra*, the anther, when ripe, perforates the septa (anther wall).

1 *Tremandra stelligera* **05460875**

L. *stelliger* = star-bearing; having stellate hairs

A very hairy, erect and spreading shrub to about 1 m. This is a common understorey shrub in moist forest or woodland habitat through the lower south-west between Bunbury and Albany. This is a plant that will be familiar to many people, but its name may not be. The flowers, though attractive, are usually few and tend to remain hidden under the leaves. The leaves turn a beautiful red as they age.

Flowers: most of the year, but mainly Sept.–Dec.

2 *Tremandra diffusa* **05515386**

L. *diffusus* = spread out; loosely or widely spreading

This prostrate little ground-cover, though inconspicuous, often covers a considerable area. Its tiny pinkish-white flowers are dotted sparsely along the stems. This is a plant that you might not realise existed until it was pointed out to you, but once identified seems to pop up everywhere! It is common in the Jarrah forest from Pinjarra to Albany.

Flowers: Sept.–Dec.

Tetratheca

G. *tetra* = four + *theca* = box; refers to the four cells of the anthers

Tetrathecas are small, heath-like plants that usually bear prolific, bright pink flowers.

3 *Tetratheca hirsuta*, Black-eyed Susan **05460956**

L. *hirsutus* = shaggy, having long hairs

A small erect but straggly shrub to about 50 cm with hairy stems and variable leaves that are sometimes opposite or in whorls of three. Often on lateritic gravelly soils in Jarrah forest. This is found on the eastern edge of our area, but it is a common species between Moora and Albany.

Flowers: Sept.–Nov.

4 *Tetratheca setigera* **05316367**

L. *seta* = bristle, stiff hair + *gero* = to bear

Also very hairy and similar-looking to the previous, but its leaves have toothed margins. Again, it is a small slender shrub to 50 cm high growing on sandy or lateritic soils in forest or open woodland. It is the more common species here and is found between Perth and Albany.

Flowers: Aug.–Nov.

P.Negus

Polygalaceae, Milkworts

A widespread family containing approximately 900 species worldwide with about 40 in Australia. The flowers are pea-like, but without the large standard petal. The two wings, actually sepals, are dominant and the flowers are generally crowded on long erect spikes. They are called Milkworts because of the milky sap that exudes from the ends of broken stems.

Comesperma

G. *come* = hair + *sperma* = seed; seeds of most species have a coma, or tuft of hairs

1 *Comesperma calymega*, Blue-spike Milkwort **05797845**

L. *calyx* = calyx + G. *megas* = large (derivation doubtful)

This is a lovely little erect, perennial herb to just 50 cm high growing on sand in heathland and open woodland. There are usually only one or two individual flowers open on the spike at one time. It is widely distributed between Kalbarri and Esperance.

Flowers: Nov.–Dec.

2 *Comesperma ciliatum* **05316235**

L. *cilium* = eye-lash; ciliate, fringed with fine hairs

A vigorous twining shrub or climber, its many stems twining tightly round each other. It grows on sandy, clay or lateritic soils in Jarrah forest or woodland between Jurien Bay and Esperance.

Flowers: Sept.–Dec.

3 *Comesperma confertum* **05316278**

L. *confertus* = crowded

A slender shrub to 1 m with flowers on a dense, pyramidal spike up to 12 cm long. The flower spikes often rise prominently above the general level of the bush. Generally grows on sandy, limestone soils, or sometimes granite, in Jarrah forest, woodland or heath. It has a mainly coastal distribution from the Arrowsmith River to Cape Arid.

Flowers: Oct.–Jan.

Note: A very similar looking species, *C. virgatum*, may also be in the area, but has a more inland distribution.

4 *Comesperma flavum* **05795257**

L. *flavus* = yellow

A delicate, perennial, multi-stemmed herb less than 50 cm high. This is a late-flowering species growing in winter-wet sand-flats and swampy areas around the west and south coasts from Kalbarri to Albany and Esperance.

Flowers: Nov.–Jan.

P.Negus

Sapindaceae

This is a large, mainly tropical family of trees, shrubs and climbers that includes such important food plants as the lychee and rambutan.

Dodonaea, Hop-bushes

After Rembert Dodoens (1517-85), Flemish physician and botanist

This is the largest Australian genus in the family, with approximately 70 species, of which 30 occur in WA. They are noted for their colourful, often winged, fruits. They are mostly found in the drier regions, but we have two here with prominent, reddish fruit. Both have small inconspicuous flowers and are dioecious with male and female flowers on separate bushes. However, they are worth looking for if you are walking in the bush in May or June. The male flowers have fat red anthers and are quite easy to find as they attract a bevy of noisy bees that gather the pollen.

1 ***Dodonaea aptera***, Coastal Hop-bush **05750156**

G. *a-* = without + *pteron* = wing

This spreading shrub can grow to about 3.5 m though it is generally lower and is found in tall heath on the coastal limestone. It is difficult to pick out amongst the other coastal shrubs except when flowering and fruiting. Its fruits, borne on long stalks, are greenish-brown when mature and are four-angled and ridged. It occurs along the west coast between Geraldton and Augusta. The female plant is shown. The male is similar to that of *D. ceratocarpa*.

Flowers: May–July; Mature fruit: Aug.–Dec.

2 ***Dodonaea ceratocarpa*** **05145481**

G. *ceras* = horn + *carpos* = fruit

Another large, spreading shrub to about 2.5 m distinguishable from *D. aptera* by its clustered, purple-brown, horned fruit capsules. It grows on laterite and granite rather than limestone and its range extends much further inland. Mostly found between Perth and Bremer Bay, but also extends along the coast to Esperance.

Flowers: May–July; Mature fruit: Sept.–Dec.

Stackhousiaceae

Stackhousia

After John Stackhouse (1742-1819), English botanist who studied seaweeds

Stackhousia is a small genus with mostly white or cream flowers that have a sweet perfume. This grows stronger at night, attracting night-flying moths that are the pollinators of these flowers.

3 ***Stackhousia monogyna***, Native Hyacinth, White Candles **05335825**

G. *monos* = single + *gyne* = woman; the flowers are arranged singly on the spike

This is a variable species, but is always a tufted perennial herb not much more than 50 cm high with white or yellow flowers and more or less leafy. It is widely distributed throughout southern Australia, in Jarrah forest, woodland or heath and grows on most soil types.

Flowers: Aug.–Nov.

Tripterococcus

G. *treis* = three + *pteron* = wing; refers to the 3-winged coccus, or carpel of the fruit

4 ***Tripterococcus brunonis***, Yellow Candles **05335817**

G. *brunonis* After Robert Brown (1773-1858), naturalist on Matthew Flinders' expedition and the most distinguished early collector of WA flora

A single stemmed slender perennial herb to about 80 cm with yellow flowers usually found on sandy or gravelly soils in Jarrah forest, woodland or heath. Widespread from Wubin in the north to Albany. The 3-winged, yellow-green fruits of this plant are perhaps as attractive as the flowers.

Flowers: Sept.–Nov.

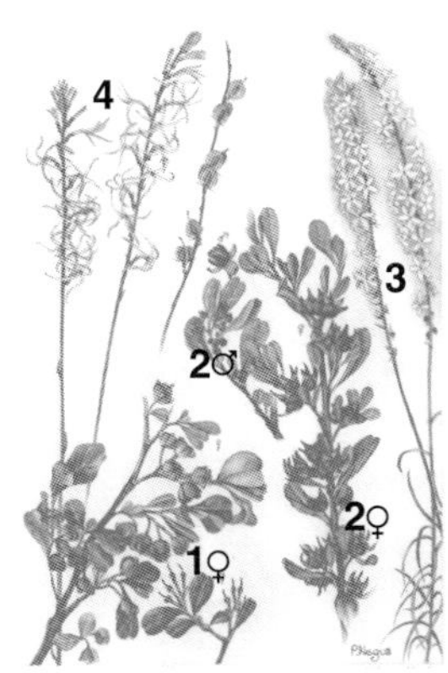

♂
♀
♀
♀
P.Negus

Rhamnaceae

A cosmopolitan family with around 900 species distributed worldwide. There are approximately 100 in Australia. They range from large trees to small heath-like shrubs and the flowers are often borne on showy heads or panicles.

Trymalium

G. *trymalia* = the eye of a needle; refers to the three slits in the fruit when it opens

1 *Trymalium floribundum* subsp. ***trifidum***, Karri Hazel, Soapbush **05460883**

L. *flos* = flower + *abundus* = abounding in

This is one of the most important understorey species in the Karri forest, but it also occurs in Jarrah forest and woodlands. It grows on a varity of soils. It is a tall shrub or a tree to 9 m high with broad, soft hairy leaves that survive the summer heat in the moist shade of the southern forests. The leaves become soapy when rubbed hard with a little water. The creamy flowers hang in large, loose panicles or clusters and have a strong scent that not everyone finds attractive. This sub-species is found in the wetter south-west extending east to Mt. Manypeaks.

Flowers: July–Nov.

2 *Trymalium ledifolium* var. ***rosmarinifolium*** **05623936**

From *Ledum* + L. *folium* = leaf

This is a much smaller shrub than the previous one, usually only growing to about 1.5 m although it can reach 2.5 m. It has narrower, shiny, dark green leaves that are lighter underneath and has less conspicuous flower panicles. It is a common shrub in Jarrah forest growing in granite or laterite areas. It is found from Dongara to Albany.

Flowers: July–Nov.

Spyridium

G. *spyridion* = a small basket; the flower heads are surrounded by basket-like leafy bracts

3 *Spyridium globulosum*, Basket Bush **05145570**

L. *globulus* = a little ball + *-osus* = abounding in

Sometimes it is the scientific name that gives meaning to the common one! This is another large bushy shrub, often growing wider than it is tall, but it can reach 5 m in height. Its dense heads of greenish-cream flowers also suggest the name of cauliflower bush to some people. It is found on the limestone sands of the coastal heath and woodlands and is widespread from Geraldton right around to the Great Australian Bight.

Flowers: July–Oct.

Cryptandra

G. *cryptos* = hidden + *andros* = man (anther); the anthers are hidden by the hooded petals

4 *Cryptandra arbutiflora* var. ***tubulosa***, Waxy Cryptandra **05750164**

From the genus *Arbutus* + G. *floris* = flower; the flowers look like those of the strawberry tree, *Arbutus unedo*

This small heath-like shrub grows to about 1 m and has masses of small tubular white flowers. It is found on sandy soils over limestone or granite, in open Jarrah woodland or coastal heath and occurs between Perth and Northcliffe.

Flowers: July–Sept.

P.Negus

Myrtaceae

The Myrtaceae is arguably Australia's most important plant family, containing the eucalypts or 'gum trees' that comprise the country's dominant forest species as well as the bottlebrushes, paperbarks, teatrees and many others. Worldwide there are at least 3,500 species in the family, mostly in the southern hemisphere, with 1500 in Australia and over half of these in WA. All contain oils in their leaves to a greater or lesser extent and many are important economically for their medicinal or aromatic properties as well as for timber and honey production. The flowers of many species are spectacular and are widely cultivated for cut flowers. There is considerable variation in the flowers with the petals providing most of the colour in some genera and the stamens doing so in others.

Astartea

Astarte was a goddess of the Phoenicians and the equivalent of Venus, the Roman god of love; the myrtles were sacred to Venus.

Astartea, like *Thryptomene*, is not a particularly spectacular genus, but contains some attractive species though they are often difficult to classify. Many were formerly included in *A. fascicularis*, but are now recognised, like these two, as distinct species. They are not yet formally named.

1 ***Astartea*** **sp.** ***Scott River (D. Backshall 88233)*** △ **05577594**

This is a slender open shrub between 50 cm and 1 m high with pinkish-white flowers. It grows on sand in winter-wet flats in the Forest Grove area and maybe elsewhere in our area. It is one of a suite of summer-flowering plants appearing in the swamps as they dry out and is found in wetlands south to Northcliffe.

Flowers: Jan.–Mar.

2 ***Astartea*** **sp.** ***juniperina (G.J. Keighery 9558)*** **05577578**

This is a larger and much bushier shrub to over 1.5 m tall, but with similar pinkish-white flowers. It grows in similar habitat and occurs between Cataby and the south coast.

Flowers: Jan.–Mar.

Hypocalymma

G. *hypo* = under + *calymma* = covering, hood; refers to the cap of the calyx in bud

3 ***Hypocalymma angustifolium***, White Myrtle **05460786**

L. *angustus* = narrow + *folium* = leaf

An erect, compact shrub to about 1 m high. Inconspicuous for most of the year this plant bursts into view when it begins to flower in late winter. The flowers may be completely white, or tinged with varying shades of pink, often deepening in colour with age. It grows on sand or clay, often in winter-wet depressions and along watercourses, but is also found on gravelly hillsides. It is common between Northampton and Albany.

Flowers: Aug.–Oct.

4 ***Hypocalymma robustum***, Swan River Myrtle **05461049**

L. *robustus* = firm, robust

A spreading more open shrub to 1.5 m high with rather larger deep pink flowers. This is one of our most vibrant spring wildflowers and it is common, though usually scattered, on sand or gravel in Banksia woodland or Jarrah forest between Perth and Albany.

Flowers: Sept.–Nov.

Calytrix

G. *calyx* = cup, calyx + *thrix, trichos* = hair; refers to the stiff hairs (awns) at the ends of the calyx-lobes

5 ***Calytrix flavescens***, Summer Starflower **05917522**

L. *flavescens* = pale yellow

This small delicate shrub growing to about 50 cm occurs just in the north of our area in Banksia woodland on sandy soil. It is widely distributed through the drier woodlands of the south-west from Eneabba to Ongerup.

Flowers: Dec.–Jan.

5 ***Calytrix flavescens***, Summer Starflower

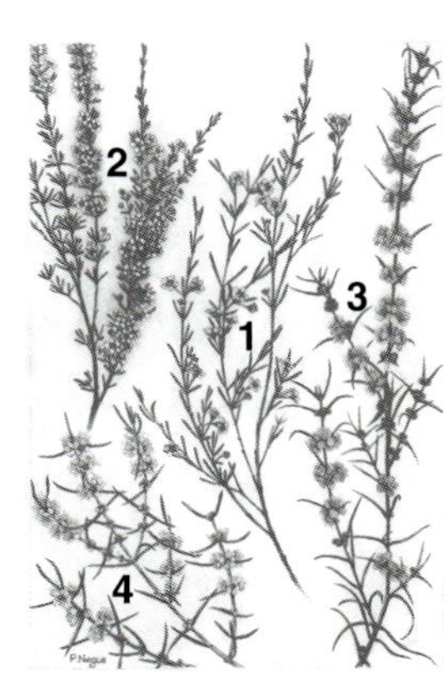

P.Negus

Agonis, Peppermints

G. *agon* = a gathering or collection; probably referring to the tight, globular flower heads of species in this genus

1 *Agonis flexuosa* var. ***flexuosa***, Peppermint **05797497**

L. *flexuosus* = full of turns; a zig-zag stem, changing directions at the nodes

One of our most common and widespread coastal species, peppermint occurs as a bushy shrub of 50 cm or less on exposed coastal heath, but grows into an elegant understorey tree up to 10 m or more in shaded Karri forest. Old trees may be gnarled and twisted into fascinating shapes. Peppermint is a significant member of most habitats: heath, woodland and forest. It grows on sandy soils mainly over limestone, but sometimes on laterite or granite and is found around the coastal plain from Perth to Bremer Bay.

Flowers: Oct.–Dec.

2 *Agonis linearifolia*, Swamp Peppermint **05577586**

L. *linearis* = linear + *folium* = leaf

Another common species, this is often the dominant shrub in our swamps and creek lines, but it also grows around the base of granite outcrops where it benefits from increased run-off. It is a medium spreading shrub up to 4 m, flowering most of the year, but particularly in spring when the masses of white flower heads resemble a dusting of snow. It occurs from just north of Perth to the south coast and east to Cape Arid.

Flowers: Oct.–Jan.

3 *Agonis parviceps*, Fine Teatree **05145651**

L. *parvus* = small + *caput* = head; the small, rounded flower heads

This is another swamp-lover, common on clayey flats and along creek-lines in the damper parts of the Jarrah forest. It is again a medium shrub to about 3 m and from a distance looks similar to the previous species, but has much shorter clustered leaves and slightly smaller flowers. The main flowering period is also earlier. It is a plant of the lower south-west, found from Bunbury to Albany.

Flowers: July–Oct.

4 *Agonis juniperina*, Wattie **05594979**

Genus *Juniper* + L. *-inus* = like

This is a tall tree to 12 m. Otherwise the flowers and leaves are similar to other *Agonis* species, especially in its immature shrubby form. The give-away is its flowering time that is much later than the other peppermints, from summer to early autumn. Again, it is found on swampy sand-flats and along watercourses in the wetter south-west from Forest Grove to east of Albany.

Flowers: Dec.–Mar.

Homalospermum

G. *homalos* = even, level, smooth + *sperma* = seed

This genus is monotypic, having only one species.

5 *Homalospermum firmum* **05624061**

L. *firmus* = firm, strong

This is a teatree with large showy flowers growing on peaty sand in winter-wet depressions and swamps. It is a large shrub growing to about 4 m. Found mostly in the lower south-west from the Capes through to Albany, with a few outliers further north as far as Perth.

Flowers: Aug.–Oct.

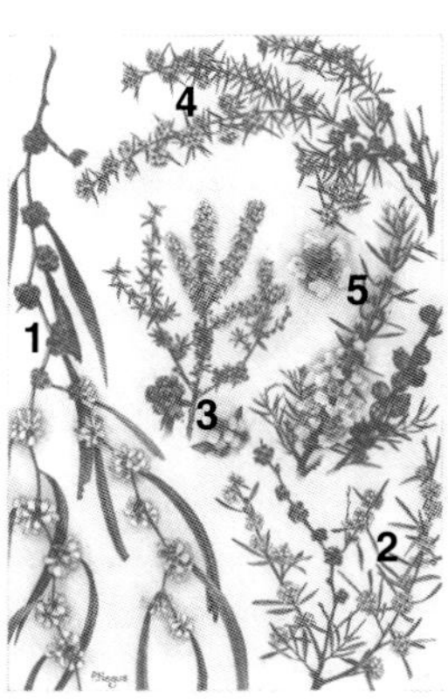

P.Negus

Darwinia

After Erasmus Darwin (1731-1802), English physician and poet, grandfather to Charles Darwin

This genus is endemic to Australia and most species are found only in the south-west. They are dwarf to medium-sized shrubs with flowers in terminal heads. The most obvious parts of many species are their colourful bell-like bracts that surround the flowers. Many *Darwinia* species have very localised distributions.

1 ***Darwinia citriodora***, Lemon-scented Darwinia **05474493**

L. *citrus* = citrus tree + *odorus* = fragrant; lemon-scented

This is one of the more widespread and variable species of *Darwinia* and is common on granite outcrops and laterite from Bindoon to Albany. It is sometimes prostrate, but can be an erect shrub up to 1.5 m and has beautiful yellow, green and red bell-flowers. It is widely cultivated.

Flowers: Aug.–Oct.

2 ***Darwinia thymoides*** subsp. ***thymoides*** **05771994**

From Thymus, thyme + *-oides* = like

This is a small matt-forming shrub only 20 cm high with green flowers and needle-like leaves found on sandy soil on granite outcrops. I have only found this plant in one spot just north of Augusta, but it may well be present on other granite outcrops in the area. It is more commonly found further north in the Darling Range east of Perth.

Flowers: Oct.–Dec.

Verticordia, Featherflowers

The 'turner of hearts', so named because the flower was thought so beautiful

3 ***Verticordia plumosa*** var. ***plumosa***, Plumed Featherflower **05624053**

L. *plumosa* = plumed, covered in feathers

A lovely and widespread species with many variants occurring between Jurien Bay and the south coast. It is a low-growing spreading shrub usually less than 1 m high that is covered with masses of pale, pinkish-purple flowers in late spring. This variant is common on clay or gravel soils on granite outcrops south of Perth.

Flowers: Nov.–Jan.

Thryptomene

G. = broken, made small; probably refers to the small flowers

4 ***Thryptomene saxicola***, Rock Thryptomene **05626919**

L. *saxum* = rock, stone + *-cola* = inhabitant

This dense shrub can grow to over 1 m, but is mostly found here as a low prostrate bush hugging the exposed coastal granite rocks on which it grows. It is found from the Capes Coast along the south coast as far as Esperance.

Flowers: June–Sept.

Pericalymma

G. *peri* = around + *calymma* = covering, hood; the leaves grade into bracts that cover the buds

5 ***Pericalymma ellipticum*** var. ***ellipticum***, Swamp Teatree **05797632**

G. *ellipticos* = ellipse; having acute rather than rounded ends (the leaves)

A tough woody shrub that can grow to 3 m, but may be quite stunted. The flowers vary from white to a delicate pale pink. Grows on clayey sand, on elevated areas within winter-wet swamps between Jurien Bay and Hopetoun.

Flowers: Nov.–Jan.

2 ***Darwinia thymoides*** subsp. ***thymoides***

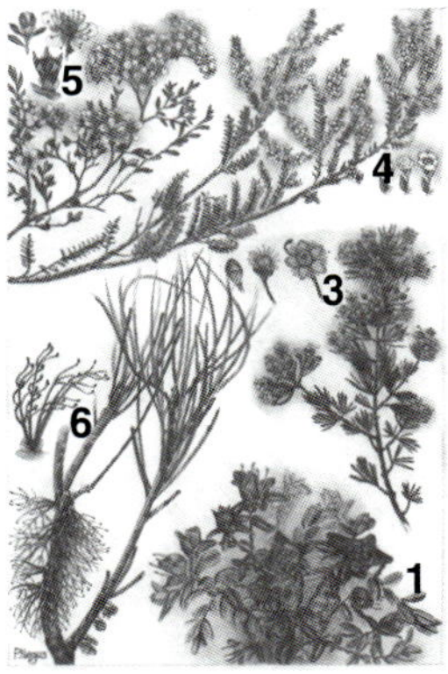

P.Negus

Kunzea

After Gustav Kunze (1793-1851), professor of botany at Leipzig

We have several *Kunzea* species in the area, but only three are shown here. They are often difficult to distinguish from each other. They are all dense medium to tall shrubs with clusters of white, or pale to dark pink flowers, except the yellow *K. spathulata.* To confuse things further, *Kunzea* species are known to hybridise readily with one another.

1 *Kunzea ciliata* **05577519**

L. *cilium* = eye-lash; ciliate, fringed with fine hairs (the bracts)

This is a dense spreading shrub to about 2.5 m growing on sandy soils on granite outcrops. It was only recently named and has so far only been collected from the Capes Coast, but it is common on the heathland of the granite cliffs around Gracetown and further north. Its flowers are generally pale pink.

Flowers: Oct.–Feb.

2 *Kunzea recurva* **05316170**

L. *recurvus* = curved backwards (the sepals)

Similar to the previous species, but with broader leaves and usually with deeper pink flowers. It also has an earlier flowering period. It is found in wetter areas on a variety of soils, but usually in winter-wet depressions and swampy country. Occurs from Mandurah to Cape Riche.

Flowers: Aug.–Nov.

3 *Kunzea spathulata* **05316197**

L. *spathulatus* = spoon-shaped

This is a taller species up to 4.5 m high with long flexible stems and lemon-yellow flowers. The needle leaves are soft and have a strong kerosene smell. It again likes to grow on damp peaty sand-flats and is found between Forest Grove and Augusta and in the Pemberton–Northcliffe area. It is similar to spearwood, *K. ericifolia,* a well-known species around Perth.

Flowers: Oct.–Nov.

Beaufortia

After Mary, Duchess of Beaufort (1630-1714), who maintained botanic gardens at Badminton and Chelsea

4 *Beaufortia sparsa*, Swamp Bottlebrush **05594863**

L. *sparsus* = scattered; refers to the leaves

The brilliant orange-red flower heads of *Beaufortia sparsa* make a spectacular sight all through the southern swamps in late summer. This shrub stands erect and tall to about 3 m among the rushes and sedges. Margaret River is about the northern limit of this bottlebrush. It is found along the south coast to Albany.

Flowers: Jan.–April.

Calothamnus, Clawflowers

G. *calos* = beautiful + *thamnos* = bush

In *Calothamnus* the stamens are prominent and are joined at their bases into distinct bundles.

5 *Calothamnus sanguineus*, Silky-leaved Bloodflower **05611938**

L. *sanguineus* = blood-coloured

An open spreading shrub that can grow to 2 m. It is found on sand over laterite or granite, but in our area it is generally wind-pruned to a low dense bush on the coastal cliffs where it mainly grows. The soft pine-like leaves are crowded around the stems. This species is widely distributed through the south-west from Kalbarri to Bremer Bay.

Flowers: Mar.–July.

6 *Calothamnus lateralis* (illustrated previous page) **05797802**

L. *lateralis* = at the side; the flowers are all on one side of the branch

This is a rather strange shrub, open and straggly, growing to about 1 m with flowers emerging from swollen branches. The new stems and leaves are a beautiful rusty red, but are prone to gall attack that causes them to become distorted. It is found in peaty swamp country and winter-wet depressions from Perth south to Bremer Bay.

Flowers: Oct.–Dec.

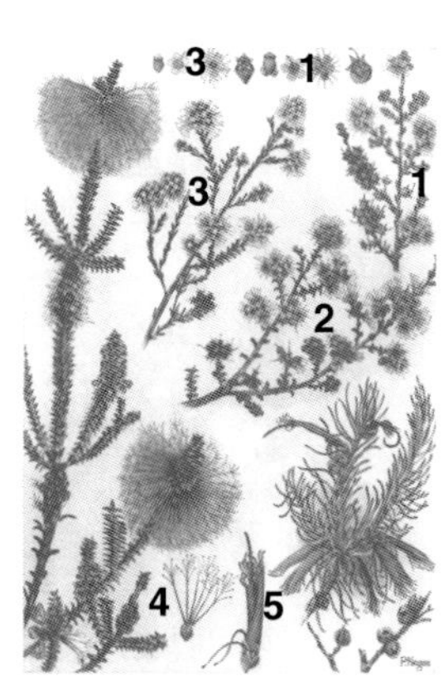

P.Negus

Melaleuca

G. *melas* = black + *leucos* = white; some Asian species have black trunks and white branches

After *Eucalyptus*, *Melaleuca* is the largest genus of the Myrtaceae. Its species range from bottlebrushes and teatrees to honeymyrtles and paperbarks.

1 *Melaleuca systena* (formerly ***M. acerosa***), Coastal Honeymyrtle **05145589**

A bushy shrub up to 2 m high, but usually wind-pruned to quite a bit less than this on the coastal limestone heath where it grows. The pom-pom lemon-yellow flowers are attractive, but not prolific. Its leaves can be soft and silky, or quite stiff. It is found around the coast between Dongara and Albany.

Flowers: Aug.–Dec.

2 *Melaleuca incana* subsp. ***incana***, Grey Honeymyrtle **05461065**

L. *incanus* = grey, hoary (foliage)

The presence of *Melaleuca incana* usually indicates swampy flats where it likes to grow in dense thickets. The small cylindrical flower heads may be white, cream or yellow, often with brownish buds as shown here. It occurs naturally from the Moore River to Albany. Like many *Melaleuca* species it grows well in cultivation and is a useful plant for damp corners.

Flowers: July–Nov.

3 *Melaleuca thymoides* **05515289**

L. *thymum* = thyme + *-oides* = like

The name presumably refers to the leaves, but the plant itself does not seem very 'thyme-like' being a stout, spreading, medium-sized shrub up to 2 m high! It grows on sandy soil over limestone, laterite, or granite, in heath or open Jarrah woodland and occurs from Perth through the lower south-west and all along the south coast to Israelite Bay.

Flowers: Sept.–Dec.

4 *Melaleuca lateritia*, Robin Redbreast **05594952**

L. *lateritius* = made of bricks, brick-red

This shrub can be easily mistaken for *Beaufortia sparsa* from a distance, growing, as it often does, in much the same swampy habitat. It has similar brilliant red flowers, but its leaves are more linear and its stamens are not branched. It also flowers a little earlier. Found from Geraldton south to Albany.

Flowers: Dec.–Mar.

5 *Melaleuca basicephala* △ **05797586**

G. *basis* = base + *cephale* = head; the flowers grow on old wood, but form the base of leafy lateral branches

This is a small woody, often straggly, plant less than 1 m high, having gorgeous mauve flowers. It is not common and is difficult to find, growing on peaty sand and clay in winter-wet swamps between Margaret River and Northcliffe.

Flowers: Dec.–Jan.

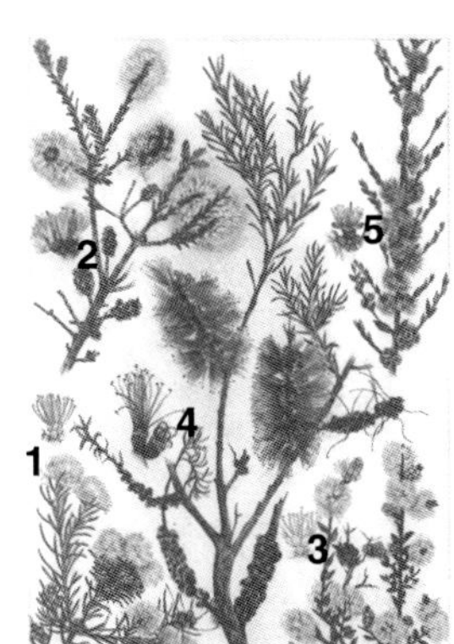

P. Negus

1 *Melaleuca huegelii* subsp. *huegelii*, Chenille Honeymyrtle **05797659**

After Carl Alexander Anselm, Baron von Hügel (1796-1870), Austrian traveller and naturalist

The common name 'chenille' is French for 'hairy caterpillar'!

This large spreading shrub or small tree can grow to 5 m tall though it is usually wind-pruned to 1 or 2 m. It is one of the dominant shrubs of the coastal limestone, often forming impenetrable thickets. Its white flower spikes with their pink buds make a beautiful show along the coastal cliffs in early summer. It is found right down the west coast from Dirk Hartog Island to Augusta.

Flowers: Nov.–Feb.

2 *Melaleuca lanceolata*, Rottnest Teatree, Moonah **05611814**

L. *lanceolatus* = shaped like a lance-head

Another very typical shrub or tree of the coastal limestone, *M. lanceolata* grows taller than *M. huegelii,* up to 8 m, often hugging the cliffs in sheltered groves. However, it is often seen pruned to a low shrub and can easily be confused with the latter. *M. lanceolata* has longer leaves, cream to pale yellow flowers that appear later in summer, and the nuts are grouped in more open clusters. This tree is common in all states and different forms are found all around the coast and inland to Coolgardie.

Flowers: Jan.–Mar.

3 *Melaleuca preissiana*, Stout Paperbark **05795249**

After Johann August Ludwig Preiss (1811-83), who collected over 2,700 species from WA between 1839 and '41

The paperbarks are another group that can be confusing. This is the one that grows furthest back from water in winter-wet depressions and around swamp edges where it only occasionally has 'wet feet'. It is a lovely shaped tree with thick sinuous branches and tight clumps of bright green foliage. It grows to 9 m on peaty, sandy soils and occurs from Eneabba to the south coast.

Flowers: Jan.–Mar.

4 *Melaleuca rhaphiophylla*, Freshwater or Swamp Paperbark **05772036**

G. *rhaphis* = needle + *phyllon* = leaf

This tree can be found growing in fresh or saline water though it is less well-adapted to saline conditions. It grows to 10 m on sandy or clay soils and, apart from its wetter habitat, can be distinguished from *M. preissiana* by its longer needle-like leaves. It also flowers a little earlier. It has quite an extensive range from Kalbarri to Fitzgerald National Park.

Flowers: Nov.–Jan.

5 *Melaleuca cuticularis*, Saltwater Paperbark **05771919**

L. *cuticula* = thin, external skin; refers to the deciduous, papery bark

This is a smaller tree than the other two paperbarks, growing to about 7 m. It tolerates salt water and salty winds, occurring from the Swan River all around the south coast where it is common in the numerous inlets and estuaries. It is found around the edge of Hardy Inlet at Augusta growing around the main estuary, while *M. rhaphiophylla* is found further back along the side creeks and streams entering the Inlet. *M. cuticularis* has distinctive star-shaped fruits and opposite leaves.

Flowers: Oct.–Dec.

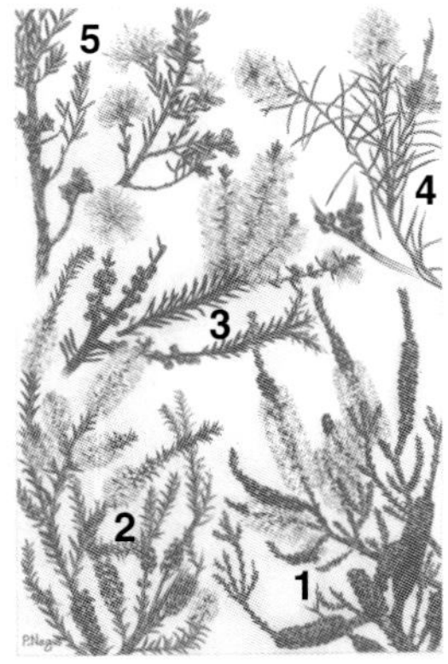

P.Negus

Eucalyptus

G. *eu* = well + *calyptos* = covered; refers to the operculum, or bud cap covering the stamens

Whenever people think of Australian plants, Eucalypts, or gum trees, are likely to be the first to spring to mind. *Eucalyptus* is one of our most important genera with over 500 species, most of which are endemic. All are trees, ranging from low, spreading mallees to the majestic tall Karri and Tingles of the southern forests. The stamens are the prominent part of the flowers and the fruits or nuts of each species are distinctive.

1 ***Eucalyptus marginata***, Jarrah

L. *marginatus* = having a border; the leaf margins are thickened

Jarrah can grow to 30 m tall, though it is difficult to find large specimens now. Together with Marri (see below) it is the major forest tree in the area. It can be recognised by its small round nuts and stringy, fibrous bark that often twists up the trunk. Jarrah and Marri are generally found growing together, with Jarrah dominant on gravelly, lateritic soils and Marri on deeper sands. Jarrah is unusual in having its main flush of growth in summer when most of our plants are dormant. Its main distribution is between Perth and Albany, but isolated pockets are found as far north as Mt. Lesueur and into the eastern wheatbelt.

Flowers: Sept.–Dec.

1 *Eucalyptus marginata*, Jarrah

2 *Eucalyptus patens*, Yarri

2 ***Eucalyptus patens***, Swan River Blackbutt, Yarri

05626692

L. *patens* = open, spreading

Blackbutt is a tall tree that can reach 45 m. It has a scattered distribution, but is generally found in valley sites where there is deep, rich soil. It is rather difficult to distinguish from Jarrah as its nuts are similar, though they have 4 or 5 rather than 3 seed capsules. However, its bark is much more like Marri, being rough and chunky. Its grey-green leaves are the same colour on both sides unlike those of Jarrah, Marri and Karri and it has fat, yellowish buds. Blackbutt occurs from east of Perth to Albany.

Flowers: Nov.–Feb.

Descriptions continue on page 146

P.Negus

3 *Eucalyptus diversicolor*, Karri **05626935**

L. *diversus* = different + *color* = colour; leaves are dark above, pale below

The name does not refer, as I used to think, to the wonderful range of colours in the bark when it peels in autumn. From pale grey, the bark turns to a patchwork of yellow, orange, pink and various shades of brown; glorious, especially after a shower of rain. Karri generally grows in single species stands and up to 30 or 40 m in this area, not as tall as in the main part of its range around Pemberton and Walpole where rainfall is higher and soils are deeper. There it may reach 60 m. Much of our Karri forest was logged in the 1890s and has regrown since then.

Flowers: Sept.–Feb., but sometimes also in winter

3 *Eucalyptus diversicolor*, Karri

4 *Eucalyptus megacarpa*, Bullich (left) **05750180**

G. *megas* = large + *carpos* = fruit

This species can grow tall and thin near swamps and streams where it is usually found, but in the Boranup area south of Margaret River it also occurs as a mallee with many trunks on the dry limestone ridge. It has smooth grey bark, similar to Karri, but its nuts are quite different and its flowers very large. This specimen was 5.5 cm across. It is a much smaller tree than Karri, growing only to a maximum of about 25 m, much less in its mallee form. It occurs from Perth to Albany and the Stirling Range.

Flowers: May–Aug.

Corymbia, Bloodwoods

The name refers to the arrangement of the flowers in clusters or corymbs. The genus has recently been split from *Eucalyptus*.

7 *Corymbia calophylla*, Marri, Red Gum (right)

G. *calos* = beautiful + *phyllon* = leaf

Marri is surely the most well-known of all our trees. It is easily recognised by its chunky, tessellated bark, frequently oozing resin and its large, characteristic 'honky-nuts'. This is the most widespread of our trees, growing in association with many other species and in a variety of situations from the Murchison River to the south coast east of Albany. It prefers sandy soils and can reach 40 m in height. Its flowering time is a boon to many animals that in mid to late summer have few choices on the menu.

Flowers: Feb.–April

5 *Eucalyptus cornuta*, Yate

L. *cornutus* = horned; refers to the long bud-caps

Yate tends to grow in isolated groves, often in valleys or depressions. It is a slender tree with an open canopy and has rough, shaggy bark that peels higher up the trunk to a smooth grey surface. It occurs from Busselton around the southern coast to Cape Arid.

Unlike the other eucalypts in the area this one has lovely greenish-yellow flowers.

Flowers: late winter and spring

5 *Eucalyptus cornuta*
Yate

5 *Eucalyptus cornuta*, Yate

6 *Eucalyptus calcicola*, Hamelin Bay Mallee (below)

L. *calcis* = limestone + *-cola* = an inhabitant

This little mallee is only found, as far as we know, in a few locations on the west side of the Ridge between Cape Freycinet and Cape Hamelin, though a species with close affinities has recently been found on the south coast. It grows up to 6 m, with cream-coloured flowers and attractive, ridged fruits. Small groves can be seen along Boranup Beach Road.

Flowers: April–June

Sterculiaceae

This is a large and mainly tropical family of trees and shrubs. The genera that are represented in the south-west are shrubs, often with spectacular flowers. Their calyx is enlarged and colourful, while the petals are much reduced or absent.

Thomasia

This genus commemorates a whole family! Twin brothers Pierre and Abraham Thomas and Abraham's three sons were all collectors of Swiss plants around 1750-1840.

The flowers are generally pink or purple with prominent anthers and the leaves are often covered with brownish hairs.

1 *Thomasia foliosa* **05626943**

L. *folium* = leaf + *-osus* = abounding in

This is a leafy plant and its white flowers are perhaps not as attractive as most in the genus. However, it is interesting, with its broad oak-like leaves and stems covered with masses of brown stellate hairs. It is a small shrub to about 1 m and it grows on gravelly laterite or granite soils in Jarrah forest and woodlands from the northern Darling Range to east of Esperance.

Flowers: July–Sept.

2 *Thomasia grandiflora*, Large-flowered Thomasia **05750172**

L. *grandis* = large + *floris* = flower

The flowers of this species are magnificent, especially when viewed against recently burnt bushland. It grows to about 1 m and likes gravelly, lateritic soils. Grows in Jarrah forest and along the coastal plain between Geraldton and Bremer Bay.

Flowers: Sept.–Nov.

3 *Thomasia macrocarpa*, Large-fruited Thomasia **05772117**

G. *macros* = large + *carpos* = fruit

An erect spreading shrub growing on loam, laterite or granite. The large flowers may be pink, blue, purple or white. In this area they are generally pink. On the granite cliff-tops north of Gracetown it is wind-pruned to a low ground-hugging form though it grows up to 2 m high in sheltered locations. Margaret River is about the southern limit of its range that extends north to Perth.

Flowers: Sept.–Nov.

4 *Thomasia paniculata* **05771862**

L. *panicula* = tuft, panicle of flowers

This is an open spreading shrub growing to 2 m and is found on sandy soils along sheltered creek-lines and swampy areas. Its stems and leaves are covered with rusty brown hairs. The long panicles of pink-purple flowers are very attractive. It is found from Perth to Albany, but does not extend very far inland.

Flowers: Sept.–Dec.

5 *Thomasia triphylla* **05797640**

G. *treis* = three + *phyllon* = leaf; there are two stipules at the base of each leaf-stalk that also look leaf-like

A dense low shrub growing not much more than 1 m high. It grows on sandy limestone soils on the coastal heathland and is sometimes one of the dominant species, though it has a patchy distribution. Its masses of purple flowers add another colour component to the yellows and blues of wattle and fan-flower. It ranges from Jurien Bay to Esperance.

Flowers: Sept.–Nov.

Lasiopetalum

G. *lasios* = shaggy + *petalon* = leaf (*bot.* petal); the underside of the calyx is covered with stellate hairs.

6 *Lasiopetalum floribundum* **05771870**

L. *floris* = flower + *abundus* = abounding in

This is a much larger, more open and spreading shrub than the *Thomasia* species described above, but its flowers, though numerous, are small and much less striking. It grows to 3 m and is widespread in Jarrah forest. It sometimes forms quite a prominent understorey for a few years after a fire. It occurs mostly between Perth and Albany.

Flowers: Oct.–Dec.

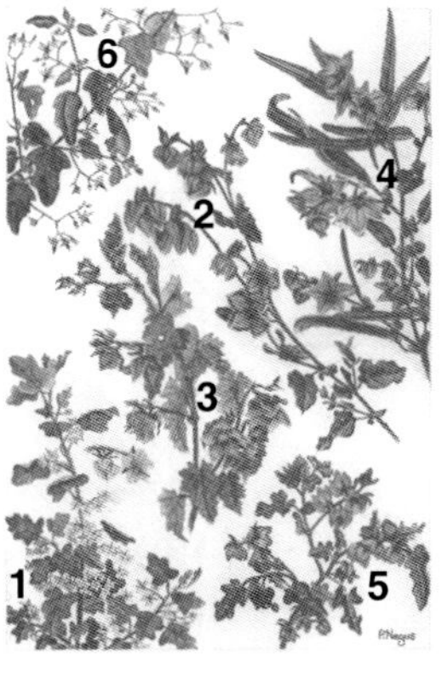

P.Negus

Thymelaeaceae

This small family includes shrubs and trees in tropical and temperate latitudes in both hemispheres. *Daphne* is a genus with which gardeners will be familiar, but in the south-west *Pimelea* is the only representative.

Pimelea

G. *pimele* = fat; may refer to the fleshy cotyledons (seed-leaves), or to the oily seeds

1 *Pimelea clavata* **05624088**

L. *clava* = a club, club-shaped

This species of Pimelea is unlike any others that grow in this area. It is a tall, open shrub up to 4 m high and has small, creamy white flower heads that are dioecious, with male and female flowers on separate shrubs. The male flower with its prominent yellow anthers is shown here. It occurs on stabilised sand-dunes over limestone, in open woodland and forest, mainly in the southern half of our area. It has flexible sinuous bark that the indigenous peoples and early settlers found useful as twine. It is found from the Capes through the southern forests to Albany and also occurs around Esperance.

Flowers: Oct.–Jan.

2 *Pimelea hispida*, Bristly Pimelea **05797543**

L. *hispidus* = rough, bristly

This small erect shrub has attractive pale pinkish flowers and grows to about 50 cm tall. It is found on sandy clay soils in winter-wet flats amongst rushes and sedges. Common in the wetter south-west from the Capes to Albany.

Flowers: Oct.–Nov.

3 *Pimelea longiflora* subsp. *longiflora* **05797578**

L. *longus* = long + *flos* = flower

A spindly open shrub to about 1 m or slightly more. Its tall creamy flower heads are often conspicuous above the general heathland in the damp sandy flats where it grows. Again found in the wetter south-west, between Bunbury and Albany.

Flowers: Oct.–Dec.

4 *Pimelea suaveolens* subsp. *suaveolens*, Scented Banjine **05625289**

L. *suaveolens* = fragrant

Another erect spindly shrub to 1 m, but this species usually has only one or two large nodding flower heads that are a lovely greenish yellow. This plant prefers a much drier habitat to the last two, growing on gravelly lateritic slopes in Jarrah forest. *P. suaveolens* is found between Jurien Bay and Albany.

Flowers: Sept.–Oct.

♂
P.Negus

1 ***Pimelea ferruginea*** **05515556**

L. *ferrugineus* = rust-coloured

When flowering this plant is one of the most spectacular sights on our coastal heathlands. Low and dense, it has a cushion-like habit that protects it from wind and storm damage on the exposed sea-cliffs where it grows. The flowers vary from pale to very deep pink and the bracts may also be pink underneath. It grows on sand over either granite or limestone, and is found from Dongara to Israelite Bay.

Flowers: Oct.–Nov.

2 ***Pimelea rosea*** subsp. ***rosea***, Rose Banjine **05316359**

L. *rosea* = rosy

Rose banjine can sometimes be confused with *P. ferruginea* when growing low and compact in heathland. In shadier situations, however, it is generally a more upright open shrub often growing quite lanky up to about 1 m high. The leaves are longer, narrower and less stiff than those of *P. ferruginea* and the flower head is on a longer stalk. It is common in woodland and Jarrah forest between Perth and Albany and grows on a variety of soils from dry sand to clay in damp depressions.

Flowers: Sept.–Dec.

3 ***Pimelea spectabilis***, Bunjong **05316286**

L. *spectabilis* = spectacular

Its large white flowers make this a prominent and eye-catching shrub of the Jarrah forest. It can grow to 2 m though it is generally less tall between 1 and 1.5 m. It has an open branching habit and usually prefers gravelly, lateritic soils and rocky slopes. It occurs from Perth to Albany and eastwards to the Fitzgerald National Park.

Flowers: Sept.–Nov.

4 ***Pimelea sylvestris*** **05577365**

L. *sylvestris* = living in woods

In our area this is the *Pimelea* of the Karri forest, growing prolifically from Boranup south to Augusta. It is a tall open shrub with slender many-branched flower heads reaching 2 m in height. The white flowers with their pinkish tinge are a lovely feature of the forest in late spring. *P. sylvestris* grows on a variety of soils. It is mostly found here on the coastal limestone though it does also grow on granite. Elsewhere, it is found in Jarrah as well as Karri forest, occurring from Bindoon to Bremer Bay.

Flowers: Oct.–Dec.

5 ***Pimelea angustifolia***, Narrow-leaved Pimelea (not illustrated) **05917492**

L. *angustus* = narrow + *folium* = leaf

A spindly open shrub with white flowers usually less than 1 m high. It is found on gravelly sandy and wet areas and on granite outcrops near Augusta. Widespread through the state.

Flowers: Nov.–Feb.

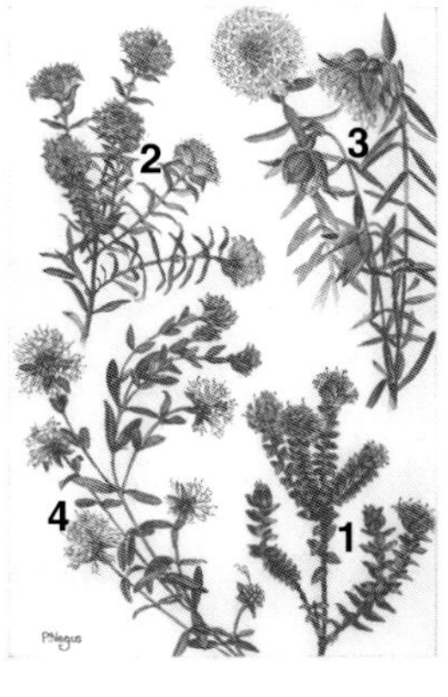

P.Negus

Epacridaceae, Southern Heaths

This is a large, southern hemisphere family closely related to the northern heaths of Europe, the Ericaceae family. There are around 200 species in Western Australia. They have spikes or clusters of tubular flowers and often have stiff pungent leaves with characteristic parallel veining. Many have succulent and edible fruits†. They generally prefer sandy, heathland environments with acidic soils.

†Note. **Never** try to eat fruits or other parts of wild plants unless you have confirmation that they are edible and not harmful!

Acrotriche

G. *acros* = at the end + *thrix, trichos* = hair; the corolla-lobes have reflexed hairs at the tips

1 ***Acrotriche cordata***, Coast Ground Berry **05750202**

L. *cordis* = heart, heart-shaped

This low dense little shrub is often overlooked, but is a common member of the heathland around the south-west and southern coasts. It grows in sand on coastal limestone cliffs and dunes. The greenish-white flowers are inconspicuous, but the white berries are easier to find. Occurs right around the coast from Leeman to the Great Australian Bight.

Flowers: Aug.–Sept.

Astroloma

G. *astron* = star + *-loma* = hem of a robe; maybe refers to the ring of hairs found inside the corolla-tube

2 ***Astroloma ciliatum***, Candle Cranberry **05145546**

L. *cilium* = eye-lash; ciliate, fringed with fine hairs

In this species in particular the red centres of the flowers do look like little stars amongst the moss-like matt of foliage of this prostrate ground cover. Candle Cranberry is found in Jarrah forest between Perth and Albany.

Flowers: July–Sept.

3 ***Astroloma pallidum***, Kick Bush **05335981**

L. *pallidus* = pale

Kick Bush is a common ground cover in Jarrah forest and woodland, only about 30 cm high with a dense matt of stiff, sharply pointed leaves that may invite a kick after causing a few pricks to the ankles! Its usually white flowers seem to have two distinct flowering periods here, early autumn and spring, but perhaps it's just that I haven't yet caught any in between. This plant has a widespread distribution between Jurien Bay and Albany, extending quite a way inland.

Flowers: Mar.–May, Sept.–Nov.

4 ***Astroloma*** sp. ***Nannup (R.D. Royce 3978)*** △ **05335957**

This is another one of those plants that, although well known in its home territory, has not yet been blessed with a specific name. Its brilliant and profuse flowering spree in autumn certainly makes it worthy of one. It is generally less than 1 m high and grows on sandy or gravelly soils in Jarrah forest between Busselton, Augusta and Nannup.

Flowers: Feb.–May

Conostephium

G. *Conos* = cone + G. *stephanos* = that which encircles; the cone-shaped corolla-tube encloses the stamens.

5 ***Conostephium pendulum***, Pearl Flower **05474477**

L. *pendulus* = hanging down (the flowers)

Easily overlooked except when flowering. Then Pearl Flower is transformed, its gorgeous, beautifully scented little tear-drop flowers hanging as purple and white pendants from the branches. It grows only to about 50 cm here, though it can reach 1.5 m. It grows in sandplain country and low ridges on sandy or gravelly soils and is found from Eneabba to Augusta.

Flowers: Sept.–Nov.

Styphelia

G. *styphelos* = tough, harsh; refers to the stiff, prickly leaves and general habit

6 ***Styphelia tenuiflora***, Common Pinheath **05457009**

L. *tenuis* = thin, narrow, delicate + *floris* = flower

This is indeed a tough, dense little shrub growing to about 1 m, usually on gravelly soils in dry Jarrah forest or woodland. The long, tubular flowers cover the plant with a profusion of white in late summer when the surrounding bush, for the most part, looks drab and grey. It is found from Bindoon to Bremer Bay, extending inland into the wheatbelt.

Flowers: Mar.–May

P. Negus

Leucopogon, Beard-heaths

G. *leucos* = white, grey + *pogon* = beard; refers to the usually densely bearded corolla-lobes

There are well over 100 species of *Leucopogon* currently recognised in the south-west and many are, as yet, unnamed. A large proportion of species are found in the far south-west corner and we have at least 12 in our area, not all of which are illustrated here. Many look superficially similar so you may be excused for finding them confusing – we do!

1 *Leucopogon australis*, Spiked Beard-heath **05456762**

L. *australis* = southern

This is an erect, bushy, medium to tall shrub up to 2 m tall. It has quite a long flower spike that may be tinged pink when in bud. It is not that common in our area, but is scattered through Jarrah forest or woodland growing on sandy or gravelly soils, often in damp situations. Found between the Moore River and Albany.

Flowers: Aug.–Oct.

2 *Leucopogon capitellatus* **05320720**

L. *capitatus* = headed + dim: *-ell-;* forming a small-head

This one is usually a small to medium shrub to about 1 m, though it is sometimes taller. It is one of the few *Leucopogon* species to have a fire-resistant rootstock. When it regenerates after a fire its stems tend to grow erect and the buds can be quite pink. As the shrub gets older it becomes much more bushy and the flowers seem to lose their pink tinge. Found in Jarrah forest on a variety of soils from Perth to Albany.

Flowers: July–Oct.

3 *Leucopogon parviflorus*, Coastal Beard-heath **05316421**

L. *parvus* = small + *floris* = flower

The flowers may be small, as with all the Beard-heaths, but there are plenty of them and the bushes are covered with their white sprays in early spring. This is a characteristic shrub of the coastal heath, growing to 2 or 3 m tall on sandy limestone soils from Dongara to Israelite Bay. It develops succulent white berries that are sweet to eat, though they have large seeds. These are much loved also by silver gulls that can sometimes be seen feasting on the ripe fruits in large numbers.

Flowers: Aug.–Oct.

4 *Leucopogon revolutus* **05320739**

L. *revolutus* = rolled back; the margins of the leaves

Very similar looking to *L. parviflorus,* but this is usually a more open shrub that has narrower leaves with rolled margins. Like *L. parviflorus*, it grows on sandy heathland in near-coastal locations also to about 2 m high. It occurs south of Geographe Bay around the south coast to Cape Arid.

Flowers: Aug.–Oct.

5 *Leucopogon verticillatus*, Tassel Flower **05145562**

L. *vertex* = revolving + dim: *-ill-;* verticillate, (leaves) arranged in a ring around the stem

At first glance this species seems vastly different from other beard-heaths, with its sparse upright stems and whorls of leaves with pendulous tassels of pink or red flowers clustered at their nodes. However, if you look closely at the individual flowers you will see that they have the same structure as other species of *Leucopogon*, though they are only sparsely bearded. The leaves are also similar, although their arrangement is different and they are much larger than those of most other species. Tassel Flower grows on lateritic or granitic soils in Jarrah and Karri forest and usually grows to about 2 m here, though in the southern Karri and Tingle forests around Walpole it can reach up to 4 m in height. Its range extends from the south coast north to Perth.

Flowers: Aug.–Oct.

1 ***Leucopogon assimilis*** **05460794**

L. *assimilis* = similar, like

This is a low bushy shrub whose name could relate to its similarity with several other species. It is generally up to 1 m high with dense, woolly flower spikes and comparatively large, stem-clasping leaves. It is found near the coast in heathland on sandy soils over granite. Its distribution is patchy from north of Perth around to Esperance.

Flowers: Aug.–Nov.

2 ***Leucopogon cordatus*** **05316456**

L. *cor, cordis* = heart; *cordatus*, heart-shaped

This lovely species is again only about 1 m high, but is densely branched. Like many *Leucopogon* species it has white flowers but pink buds that are very attractive, as are the glossy reddish-green, recurved heart-shaped leaves. It is common on sandy heathlands behind the Leeuwin-Naturaliste Ridge where, in places, it is the dominant understorey species. Found mostly between Mandurah, the south-west Capes and Denmark.

Flowers: Aug.–Oct.

3 ***Leucopogon elatior*** **05213932**

L. *elatior* = taller

This is a slender but erect, straggly shrub only growing to about 50 cm and usually taller than it is wide. It has small, rather sparse, flower spikes and is found scattered through bushland in sandplain country between Perth and Albany. Very similar to, and possibly not distinct from, *L. glabellus*.

Flowers: much of the year

4 ***Leucopogon propinquus*** **05335973**

L. *propinquus* = related to, resembling – presumably other *Leucopogon* species

This is a densely spreading and rigid shrub to about 1.5 m that has particularly pungent (sharp pointed) leaves. It is quite easy to identify by its flowering time in early autumn. Its edible, round, green berries develop through the winter. This is another common Jarrah forest species, growing on sandy or gravelly soil. It is found from Jurien Bay to Bremer Bay.

Flowers: Mar.–May

5 ***Leucopogon*** sp. ***Margaret River (J. Scott 207)*** **05611806; 05750229**

This species is very similar to *L. pendulus* (see below), except that its flowers and leaves are consistently larger and have slight differences in structure. The fruits are also larger and rounder – developing fruit of *L. pendulus* above and left of flower – and it has an earlier flowering period. It grows on gravelly soils in Jarrah forest between Capel, Nannup and Augusta.

Flowers: Mar.–May

6 ***Leucopogon pendulus*** (fruit only illustrated) **05145600; 05750210**

L. *pendulus* = hanging down, drooping (flowers)

Very similar to the species above, but has smaller flowers. It is often found on seasonally wet flats and granite outcrops through the lower south-west between Bunbury and Albany.

Flowers: July–Sept.

Note: The last two species belong to a group that may soon be re-classified into a separate genus.

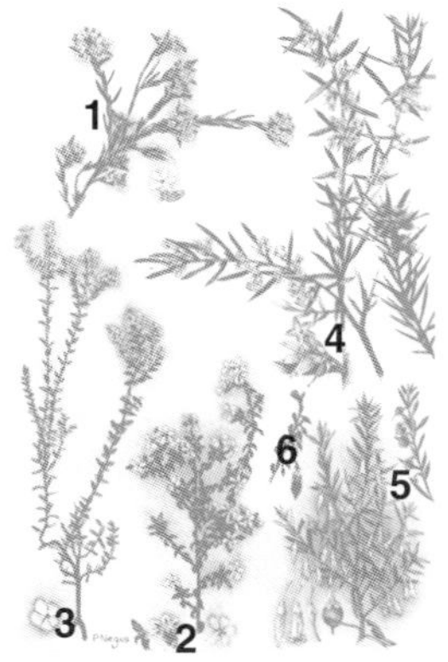

P.Negus

Andersonia

This genus name is most economic, as it commemorates three botanists in one! William Anderson (1750-78), surgeon and botanist on Captain Cook's second and third voyages; Alexander Anderson (*d* 1811), curator of the botanic gardens at St. Vincent; and another William Anderson (1766-1846), curator of the Apothecaries' Botanic Garden at Chelsea.

1 *Andersonia caerulea*, Foxtails **05145619**

L. *caeruleus* = deep sky-blue

Foxtails are beautiful little plants with dense pink and blue spikes of flowers that are a delightful surprise when found in sandy heath and woodland in the middle of winter. They are small and bushy, not usually exceeding 50 cm and are widely distributed through the lower south-west and east to Esperance. The species is very variable , and is soon to be split into several entities.

Flowers: July–Oct.

2 *Andersonia involucrata* **05772095**

L. *involucre* = a cloak to wrap around, *involucrum* = envelope; having leaves or bracts surrounding the base of the flower head

This can be a dense and compact, or straggling shrub, only to about 50 cm high with many terminal white or pinkish flower heads. It grows on sandy soils over laterite or granite, often in moist locations or on rocky breakaways. Found between Jurien Bay and Augusta.

Flowers: Nov.–Dec.

Lysinema

G. *lysis* = loosing, or setting free + *nema* = thread (bot. stamen); the stamens are free from the corolla, unlike in many other heaths.

3 *Lysinema ciliatum*, Curry Flower **05474469**

L. *cilium* = eye-lash; fringed with fine hairs – the sepals and bracts

This is a common and variable species with several different forms being recognised. It is an erect shrub up to about 1 m tall with distinctive brown and white flower spikes. Rub the flowers between your fingers and smell their rather curry-like aroma that gives this plant its common name. It grows on a variety of soils in open heath or woodland from Kalbarri to Israelite Bay.

Flowers: July–Oct.

4 *Lysinema conspicuum* **05797837**

L. *conspicuus* = striking

A small, slender plant of damp, sandy flats and winter-wet swamplands that is less than 1 m tall. Its large white flower spikes are relatively more conspicuous, hence the name. It grows along the southern coastal plains between Margaret River and Albany.

Flowers: Nov.–Jan.

Sphenotoma, Paper-heaths

G. *sphenos* = wedge + *tome* = place from which something has been cut; the lower part of the corolla below the lobes is wedge-shaped

5 *Sphenotoma capitatum* **05316391**

L. *capitatus* = headed, forming a head

This is a slender, erect little shrub to about 1 m growing on sandy or gravelly soils in heath and Jarrah forest. Each plant has several large, papery white flower heads. It is found from Pinjarra to Albany.

Flowers: Oct.–Nov.

6 *Sphenotoma gracile*, Swamp Paper-heath **05797551**

L. *gracilis* = thin, slender

A slender, straggling shrub to about 1 m with small white flowers in terminal spikes. Its stems are often supported by adjacent vegetation. It grows on swamp-edges or winter-wet depressions between Busselton and Albany.

Flowers: Oct.–Dec.

4 *Lysinema conspicuum*

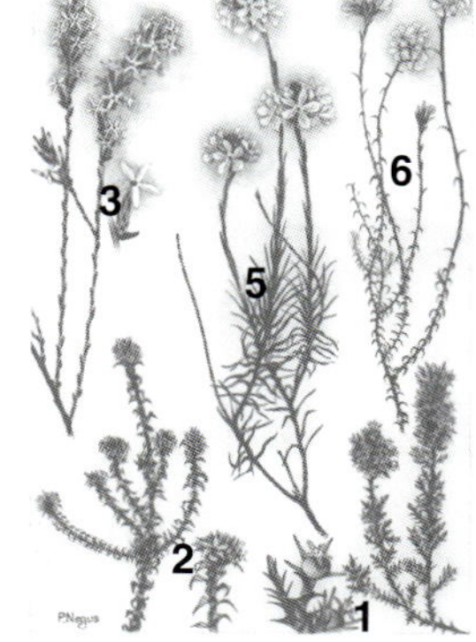

P.Negus

Primulaceae

Samolus

Pliny used this name for a marsh plant. Most species grow in wet places.

1 *Samolus repens*, Creeping Brookweed **05772435**

L. *repens* = crawling, prostrate

This pretty little plant is usually indicative of springs and seepages along the coast where it is found at the junction between granite and limestone. It is seldom more than 30 cm high. Elsewhere it is found around salt-marshes and is widely distributed throughout the south-west. It belongs to the same family as the Scarlet and Blue Pimpernels (see section on weeds).

Flowers: Oct.–Feb.

Loganiaceae

A diverse family, occurring mainly in the tropics.

Logania

After James Logan (1674-1751), born in Ireland, governor of Pennsylvania 1736

2 *Logania serpyllifolia* subsp. *angustifolia* **05316294**

L. *serpyllum,* from wild thyme, *Thymus serpyllum* + *folium* = leaf

A small shrub of Jarrah forest and woodland, this is one of those 'inconspicuous, but found everywhere' plants. It is usually less than 50 cm and its star-shaped flowers are seen in a variety of habitats. It is found from Perth to Albany and along the south coast to Esperance.

Flowers: Aug.–Dec.

3 *Logania vaginalis*, White Spray **05316383**

L. *vaginalis* = sheathed

This tall, open shrub is very different to the last, growing to 2 m or more with large, conspicuous flower heads. It prefers sandy or lateritic soils, but is widely distributed in Karri and Jarrah forest, woodland, or heath around the west and south coasts from Leeman to Esperance.

Flowers: Aug.–Oct.

Violaceae

A worldwide family of over 900 species with only a very few members in Australia.

Hybanthus

G. *hybos* = hump-backed + *anthos* = flower; refers to the pouched petal

4 *Hybanthus calycinus*, Wild Violet **05626978**

L. *calyx* + *-inus* = -like; used when the calyx is somehow unusual. Here it is quite prominent

An erect perennial herb with many flower stems growing to 50 cm or more on sandy, limestone soils. It occurs all down the west coast from Kalbarri to Margaret River, which is about its southern limit. Its little mauve clumps are very attractive along coastal paths in early spring before much else is in flower.

Flowers: July–Oct.

Lamiaceae

This is a large family whose flowers have a distinctive, two-lipped corolla. Many members also have square stems with opposite, often aromatic, leaves. They include many well-known herbs such as the mints, thyme, sage, marjoram and lavender. In Western Australia there are many more species in the dry interior, but this family is also represented in our far south-west corner.

7 *Hemigenia podalyrina*

Hemiandra

G. *hemi-* = half + *andros* = man (bot. stamen); anthers are 1-celled, the lower part being much reduced

5 *Hemiandra australis* ms ∆ **05577225**

L. *australis* = southern

This is a lovely species that is an erect shrub growing to nearly 1 m. It is found on granite outcrops near Augusta, but is not common. It has a fairly restricted distribution between Augusta and Walpole.

Flowers: Oct.–Nov.

Hemigenia

G. *hemi* = half + *genia*–may come from *geneion* = that part of the face covered by a beard – part of the anther is bearded – or *genea* = birth – only one of each pair of anther cells is fertile

6 *Hemigenia rigida* **05515408**

L. *rigidus* = stiff

This is an erect, open shrub to 60 cm or so with small violet flowers. You will find it on sandy or lateritic soils on hill slopes and granite outcrops. It is fairly inconspicuous, but quite common around Margaret River. Its range extends from north of Perth to Denmark and inland into the wheatbelt.

Flowers: Sept.–Dec.

7 *Hemigenia podalyrina* **05772079**

Genus *Podalyria* + *-inus* = like

I have, so far, only found this plant in one location near Gracetown, where it looked rather out of place! However, it has been recorded in the area before, although its main range is along the south coast from Northcliffe to Albany. It is a medium-sized, bushy shrub with attractive pale foliage and mauve flowers, growing on gravelly soil in tall heathland.

Flowers: Oct.–Nov.

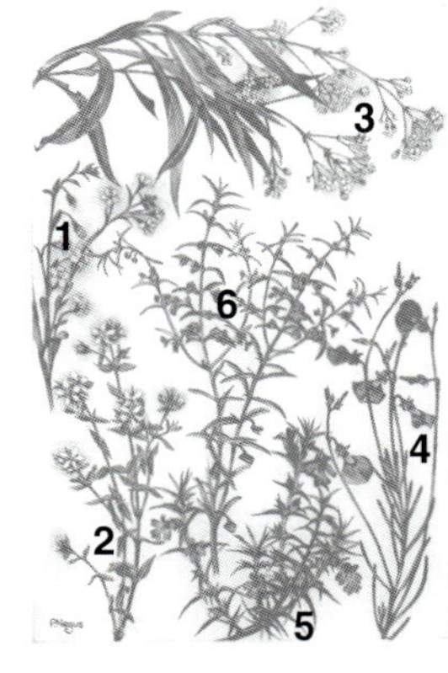

P.Negus

Apiaceae

This family used to be called Umbelliferae because the flowers are arranged on the flower heads on stalks radiating like the spokes of a wheel, or 'umbels', in other words, like an umbrella. It contains many well-known vegetables and herbs, such as carrots, celery, fennel, parsley, and most have an aromatic 'carrot-like' scent when crushed. There are a large number of species in the south-west, many seemingly inconspicuous and 'weed-like' in appearance, but they do belong here. The more species I become acquainted with the more intriguing I find them.

Platysace

G. *platys* = wide, flat, broad + *sacos* = sack, bag; refers to the broad fruit

1 *Platysace filiformis* **05577411**

L. *filum* = thread + *forma* = shape

The flattened stems of this herb are distinctive, being very similar in form to the tapeworm plant (*P. compressa)* but rather more slender. It is quite a tall plant that can grow up to 1 m when supported by other vegetation, though it is usually straggly and sprawling. It likes moist situations and is quite common here in the Karri forest. It occurs from Perth to Hopetoun, mostly fairly near to the coast.
Flowers: most of year.

2 *Platysace tenuissima* **05577462**

L. *tenuis* = thin, delicate, *-issimus* = most

This is one of those common little plants that often go unnoticed, but it is really beautiful when examined closely. It is a small herb, generally less than 50 cm high, though often quite dense. It is found in Jarrah forest or woodland on gravelly soil, or calcareous sand. Usually white or cream, the flowers can sometimes range through to a delicate yellow or pink. It is found between Perth and Walpole.
Flowers: Sept.–Jan.

Trachymene

G. *trachys* = rough + (either) *hymen* = skin, or *mene* = moon; refers to the fruit, and either would be appropriate!

3 *Trachymene coerulea* subsp. ***coerulea***, Blue Laceflower **05577284**

L. *coeruleus* = variant spelling of *caeruleus* = deep sky-blue

A stout, annual herb, growing to 2 m. This is a gorgeous plant that would look at home in any garden. It grows on granite outcrops around Cosy Corner and I found it in quantity at Cape Freycinet after a fire. This subspecies extends south from Yanchep to our area; plants growing north to Shark Bay are of a different subspecies. The flower head is a dense cluster of individual flowers that become darker towards the centre. Only the styles peep out from the petals of outer flowers, while the stamens of inner flowers are also visible.
Flowers: Nov.–Jan.

Note: We were interested to find that the little jewel beetle on the specimen illustrated here was the same as that found on plants on Rottnest Island, painted by Elizabeth Rippey for the book *Plants of the Perth Coast and Islands*.

4 *Trachymene pilosa*, Native Parsnip **05625114**

L. *pilosus* = hairy

There are several common names amongst plants in this family, reflecting the similarity of some to old-world species. What a relief it must have been to the early settlers to find at least one or two plants that resembled those back home, even if they did not make such good eating! This is an erect annual herb to only 30 cm high, common on sandy soils in forest, woodland or heath. It is widespread between Shark Bay and the Great Australian Bight.
Flowers: Aug.–Nov.

P.Negus

Xanthosia

G. *xanthos* = yellow; refers to the yellow hairs on some species

This is the genus that includes the well-known Southern Cross, *Xanthosia rotundifolia.* That species is found around the south coast and not here, but we do have some others from this group. These are small ground-cover plants and not so spectacular, but interesting nonetheless.

1 *Xanthosia candida* **05772494**

L. *candidus* = dazzling white, pure; maybe referring to the undersides of the leaves

A prostrate or upright perennial herb to only about 30 cm, with rather lovely pale grey-green foliage that is common in Jarrah forest and open woodland from Bindoon to Albany.

Flowers: Oct.–May

2 *Xanthosia tasmanica* **05772508**

tasmanicus = Tasmanian

A small ground-cover species can often be overlooked and it was quite a while before I recognised this little plant, having walked past and around it for years! It is a prostrate perennial herb less than 20 cm high with tiny greenish flowers, growing on sandy or clay soils in forest and woodland. While similar in size and form, its foliage is much greener than that of the last species. It is found between the Capes and Albany.

Flowers: Oct.–Jan.

3 *Xanthosia atkinsoniana* **05797527**

After Caroline Waring Atkinson (1834-72), botanical collector, taxidermist, botanical artist and writer

This is a small, erect perennial herb that can grow to 1 m. I have so far only found it on the western edge of the Boranup Karri forest, but it may be more widespread. It grows on sandy soil over limestone and it ranges from Moora to the south coast and inland as far as Narrogin and Katanning.

Flowers: Oct.–Nov.

Pentapeltis

G. *pente* = five + *pelte* = small shield; refers to the 5 calyx-lobes that are attached peltately, meaning that their stalks are attached to the underside rather than the edge of the calyx-lobe

4 *Pentapeltis peltigera* **05795001**

G. *pelte* (as above) + *gero* = to bear

Prostrate perennial herb only about 20 cm high, with interestingly shaped leaves. It grows in Jarrah forest and woodland and is another little plant that is easily overlooked. It occurs between Toodyay and Augusta.

Flowers: Dec.–Mar.

P.Negus

Apium

L. Name used by Pliny for parsley or celery

1 *Apium prostratum* var. ***prostratum***, Sea Celery **05795133**

L. *prostratus* + thrown to the ground, lying along the ground

Sea celery is usually a straggly, slender-stemmed plant, though it sometimes grows stout and erect. It is a biennial or perennial herb, spreading along the ground, but also reaching to about 1 m high. It is common around the south-west coast and inland between Kalbarri and Esperance. It prefers sandy, damp sites, such as creek-lines, but is also found on rocky shores. It can be eaten, but has a very salty taste.

Flowers: Nov.–Feb.

Eryngium

From G. *eryggion,* name used in ancient Greece for a type of thistle, from *eryggano* = to belch; that plant was regarded as a remedy for flatulence!

2 *Eryngium pinnatifidum* subsp. ***pinnatifidium***, Blue Devils **05515432**

L. *pinnatus* = feathered + *findo* = to split; 'pinnatifid' is used for a leaf-blade that is divided into lobes half-way to the mid-rib

This is a very striking species with its lovely blue foliage and flowers standing out in the sparse heathland on the limestone ridge where it occurs. In fact, it looks rather out of place and I am often asked if it is a weed. It is a stiff, erect perennial to about 50 cm high and is found between Geraldton and Albany.

Flowers: Nov.–Jan.

Centella

Derivation uncertain, maybe from L. *cento* = patchwork cover

3 *Centella asiatica*, Pennywort, Gotu Kola **05795028**

L. *asiaticus* = Asian

This is a fascinating plant that, unlike most WA species, is widespread through Asia and found wild all over Australia. It is a low, creeping ground-cover with tiny white or pinkish flowers, rarely more than 30 cm tall and is found in wet depressions and along water-courses. It has an extensive reputation for its healing properties and is claimed to cure a broad spectrum of ailments from skin diseases, fevers, diarrhoea, throat infections and inflammations, to arthritis. It is said to increase one's energy, concentration and memory and to enhance longevity. Advocates extol the virtues of eating a leaf a day, but beware as too much can cause headaches, dizziness and sleeplessness! It has a coastal distribution from Geraldton to Esperance.

Flowers: Oct.–April.

Convolvulaceae

Dichondra

G. *dis* = double + *chondros* = a grain; refers to the double carpels (fruits)

4 *Dichondra repens*, Kidney Weed **05626781**

L. *repens* = crawling, prostrate

A slender, creeping perennial herb no more than 20 cm high, common on sandy soil on the forest floor or under heathland shrubs on stabilised dunes. Its soft little leaves will be easily recognised, but the flowers are tiny and well-hidden and not nearly so familiar. It is found from Perth through the southern forests and around the coast as far as the Great Australian Bight.

Flowers: Sept.–Nov.

P.Negus

6 *Opercularia volubilis*, Twining Stinkweed

Menyanthaceae, Marshworts

This is a small family of mostly aquatic or semi-aquatic herbs. They are distributed worldwide, but more than 30 species are found in Australia.

Villarsia

After Dominique Villars (1745-1814), French botanist and physician

The flowers of *Villarsia* are yellow or white. Ten species are found in the south-west.

1 *Villarsia latifolia* **05335930**

L. *latus* = broad + *folium* = leaf

This is a large, robust, perennial herb growing to 1 m and is found on sand, loam or peat soils in moist places, streams and swamp edges. Here, though, it is quite prevalent through the Karri forest, often on quite dry slopes where, presumably, it is able to survive due to the high rainfall and sheltered situation. It is found between Perth and Albany, not extending very far inland.

Flowers: Oct.–Dec.

2 *Villarsia parnassifolia* **05577543**

From the genus *Parnassia* + L. *folium* = leaf

Another semi-aquatic perennial herb growing to 1 m, with yellow flowers. It is more delicate and has smaller flowers than the last species and it has a clumped habit. It is found in shallow water or damp soil in swamps, wet flats and drainage lines from Mandurah south and along the coast to Esperance.

Flowers: Nov.–Dec.

Rubiaceae

Opercularia, stinkweeds

L. *operculum* + cover, lid; refers to the lid-like valves on the fruit (see illustration)

This is a genus of diminutive perennial herbs that rarely rate a mention in wildflower books. However, there are several species and they are surprisingly common once you start looking. Four species are included here, but there may be others in the area as well. Their common name is derived from their unpleasant odour when crushed.

3 *Opercularia echinocephala*, Bristly-headed Stinkweed **05750148**

G. *echinos* = hedgehog + *cephale* = head

The name is very apt for this spreading perennial! It grows to only about 30 cm and grows on lateritic or sandy soils in Jarrah forest. Though this little plant is fairly obscure, on close examination its pink and green flowers are actually quite pretty. It grows from Perth to Augusta and along the south coast, with isolated occurrences as far east as Esperance.

Flowers: Sept.–Nov.

4 *Opercularia hispidula*, Hispid Stinkweed **05750210**

L. *hispidus* = rough, with short stiff hairs + dim *-ul-*

This is the most common of the local stinkweeds and is a familiar plant of roadside verges. Most people would probably think it was a weed, but it does belong here and is widespread throughout the south-west on sandy, gravelly or clay soils in a variety of habitats from north of Perth to Albany and along the south coast to Cape Arid.

Flowers: Sept.–Dec.

5 *Opercularia vaginata*, Dog Weed **05577551**

L. *vaginata* = sheathed; the seeds are partially enclosed

For a group of fairly inconspicuous plants *Opercularia* species have more than their share of common names. I do not know the derivation of this one. This is a straggly, spreading, perennial herb, less than 50 cm high, growing in coastal heathland on stabilised sand-dunes over limestone. It is widespread from Shark Bay to Israelite Bay.

Flowers: Aug.–Oct.

6 *Opercularia volubilis*, Twining Stinkweed **05624916**

L. *volubilis* = spinning, twisting

This is a scrambling, twining perennial that can be quite robust, reaching up to 2 m high. Apart from its twining habit, it looks very similar to *O. hispidula*, though it seems to favour dampish, forest situations. It occurs between Margaret River and Albany.

Flowers: Sept.–Nov.

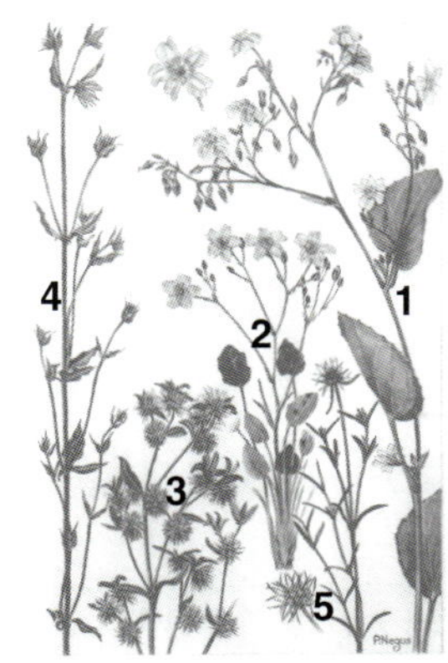

P.Negus

Solanaceae

This is a large and diverse family of plants that includes some of our most important food plants, such as potatoes, tomatoes and capsicums. However, it also includes some with a high alkaloid content that are poisonous or used as drugs and narcotics such as tobacco *Nicotiana* spp. and, in WA, Pituri, *Duboisia hopwoodii,* which the aborigines used as a narcotic and fish poison.

Solanum

Name used by Pliny for *S. nigrum,* Black Nightshade.

Most of the many species of *Solanum* in WA are found in the dry, inland regions, but we have one indigenous species here.

1 *Solanum symonii* **05611792**

After David Eric Symon (*b* 1920), botanist, Waite Agric. Research Institute, Adelaide

S. symonii is a large leafy shrub of coastal limestone and sand-dunes growing to 2 m. The bright purple flowers are often hidden amongst the dark green foliage. It is found right around the west and south coasts from Geraldton into South Australia.

Flowers: Sept.–Mar.

Anthocercis, Tailflowers

G. *anthos* = flower + *cercis* = tapering rod; refers to the narrow corolla lobes

2 *Anthocercis littorea*, Yellow Tailflower **05515475**

L. *littoreus* = pertaining to the sea-shore

A bushy, spreading shrub to 3 m tall found in coastal heathland on calcareous, sandy soils and sand-dunes. It thrives on disturbed sites and is an early coloniser after fire. It extends from Shark Bay around the coast to the Great Australian Bight.

It is difficult to distinguish from *A. ilicifolia*, which has large, leafless inflorescences and brighter flowers and may also be in the area.

Flowers: most of the year

Myoporaceae

Myoporum

G. *myo* = to be shut or closed + *poros* = pore; refers to the closed appearance of the glands on the leaves

This is a genus of shrubs or small trees with regular white or purplish flowers. The two species here both have little blue berry fruits.

3 *Myoporum insulare*, Blueberry Tree, Boobialla **05772419**

L. *insula* = island

Blueberry tree is, in fact, usually a dense shrub varying in height from a few centimetres up to 2 m. It has quite thick, succulent leaves and grows on sand-dunes and limestone in just a few coastal locations in our area, such as Gracetown. It is found around the drier parts of the west and south coasts from Dirk Hartog Island to the Nullarbor.

Flowers: Oct.–Dec.

4 *Myoporum oppositifolium*, Twin-leaf Myoporum **05577357**

L. *oppositus* = placed opposite + *folium* = leaf

This species of *Myoporum* is common both in our Karri forest and on the coast; for example, at Prevelly Park. It is a slender or sprawling shrub up to 1.5 m high with dark, shiny leaves. It grows on sandy soil over limestone and is found from the Capes coast to Albany, but not extending far inland.

Flowers: Oct.–Jan.

Scrophulariaceae

Veronica, Speedwells

Named by the Italian botanist, Pier Antonio Mattioli in 1554 in honour of St Veronica, or a nun of the same name, who was beatified in 1517.

5 *Veronica distans* **05626714**

distans = separate, different, standing apart

This soft little sky-blue herb is found through the Karri forest understorey and in adjacent heathland, generally fairly well hidden amongst other ground-cover plants. It grows about 20 cm high, creeping along in the shade of the forest floor. It has a scattered distribution between the Capes and Albany.

Flowers: Oct.–Dec.

5 *Veronica distans*

P.Negus

Lobeliaceae

This is a small family of attractive herbs. They flower late in spring and in summer and have the unusual characteristic of using the sap stored in their leaves and stems to sustain their flowers and fruit. As this happens, they wither from the base upwards.

Lobelia

After Matthias de l'Obel (1538-1616), Flemish botanist

Lobelia species can be easily recognised by the 'fleur de lys' shaped blue flower. They have milky sap in their stems, as do species of *Isotoma*, the other genus of the family found in our area.

1 ***Lobelia alata***, Angled Lobelia **05577349**

L. *alatus* = winged (stems)

Angled Lobelia is common in wet locations such as streams and spring seepages near the coast where it is often found with *Samolus repens*. It is a prostrate or ascending perennial herb, usually only to about 50 cm high, with small flowers, brightening its environs throughout the summer. It is widespread through the south-west between Kalbarri and Israelite Bay.

Flowers: Nov.–May

2 ***Lobelia gibbosa***, Tall Lobelia **05795036**

L. *gibbosus* = swollen, pouched; refers to the flower

Slender and erect, but often quite robust, this is another late-flowering annual herb not usually appearing until after Christmas. It is found in a variety of situations and on a variety of soils. The stems are often quite succulent. Again, found all through the south-west from Shark Bay to Esperance.

Flowers: Dec.–Mar

3 ***Lobelia tenuior***, Slender Lobelia **05795230**

L. *tenuior* = thinner, narrower, more delicate

This is a lovely annual herb with slender, erect stems from a few centimetres to 50 cm or more. The flowers are usually a brilliant blue, with a white or yellow throat, often forming drifts of colour with other later-flowering annuals such as *Podolepis gracilis* and Swan River Daisy, *Brachyscome iberidifolia*. It grows mainly on limestone sand amongst the coastal heath, but also on the red loam of the Karri forest. It is found from Yanchep to Albany.

Flowers: Oct.–Jan.

Isotoma

G. *isos* = equal + *tomos* = a piece cut off, slice; unlike *Lobelia*, this genus has corolla-lobes that are all nearly equal

4 ***Isotoma hypocrateriformis***, Woodbridge Poison **05515521**

G. *hypo* = under + *crater* = bowl, crater + L. *forma* = shape; saucer-shaped

A succulent, erect annual herb growing to 50 cm or more, its flowers varying in colour from white, pink and purple to blue. It is widespread throughout the south-west in a variety of habitats.

Flowers: Oct.–Jan.

5 ***Isotoma scapigera***, Long-scaped Isotome **05515440**

L. *scapus* = stem, stalk, *bot.* flower stalk, leafless or almost so, growing from the ground + *gero* = to bear

A pretty little rosetted annual herb only about 20 cm high that is conspicuous for its beautiful colour. Here it grows on sand-dunes and limestone sand in dry coastal heathland though apparently elsewhere it favours wet depressions and salt-lake edges. It has a scattered distribution between Mullewa and Cape Arid.

Flowers: Oct.–Dec.

Campanulaceae

Wahlenbergia

After Goran Wahlenberg (1780-1851), professor of botany at Uppsala

There are nine currently recognised native species and one better-known introduced one, the Cape Bluebell from South Africa, *W. capensis*.

6 ***Wahlenbergia littoricola*** **05626773**

L. *litoralis* = belonging to the sea-shore + *-cola* = inhabitant

A slender, ascending perennial to about 30 cm growing on clay or loam soils. Found scattered through the lower south-west.

Flowers: Oct.–May

6 *Wahlenbergia littoricola*

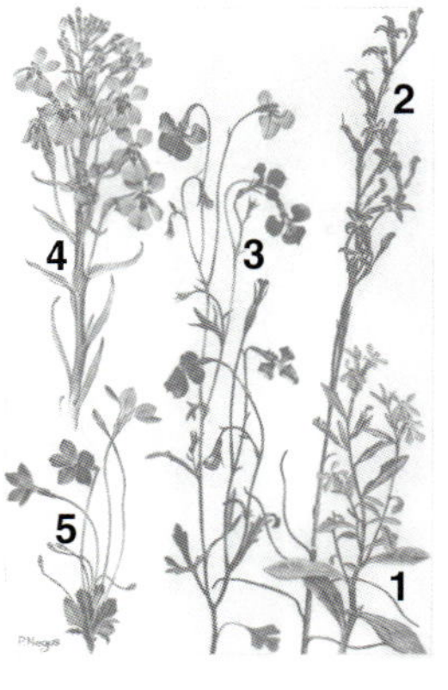

P.Negus

Goodeniaceae

This is a family of 400+ species that is almost entirely Australian, apart from a few species scattered throughout the southern hemisphere. The main family characteristic is the cup at the top of the style, called the indusium, into which the pollen is deposited. The corolla-lobes of the flowers are also winged. Members of the family are found in most habitats except rainforest.

Goodenia

After Samuel Goodenough (1743-1827), Bishop of Carlisle and first treasurer of the Linnean Society

1 *Goodenia eatoniana* **05335892**

After Henry Francis Eaton (1831-1912), under-treasurer of Victoria and strong supporter of scientific activities

A beautifully striking, late-flowering species with bright blue flowers and silvery blue-green foliage. It is mainly found on sand or clay, winter-wet flats and streamlines and grows to about 30 cm. It is found in the lower south-west from about Waroona to Albany.

Flowers: Nov.–Feb.

2 *Goodenia pusilla*, Delicate Goodenia **05577535**

L. *pusillus* = very small, insignificant, weak

I used to confuse this plant with *Velleia trinervis*, until I looked at it closely. it is a small, ascending annual herb, growing to only about 10 cm and found on peaty sand in swamps and seepage areas. It is found through the lower south-west from Gingin to Jerramungup.

Flowers: Nov.–Jan.

Velleia

After Thomas Velley (1748–1806), English botanist, specialist in seaweeds and other marine plants

3 *Velleia trinervis* **05577217**

L. *tres* = three + *nervis* = sinew, nerve

Perennial herb up to 50 cm high with a rosette of basal leaves. It varies tremendously in height and stature from small and slender to tall and robust. It is mainly found in damp locations and is common from Dongara to Cape Arid.

Flowers: Oct–Jan.

Lechenaultia

After Jean-Baptiste Leschenault de la Tour (1773–1826), botanist on Baudin's expeditions, 1800–04. He collected at Cape Naturaliste, the Swan River and Shark Bay. Robert Brown, who named the genus, originally misspelt the scientific name and it therefore remains without the 's'.

4 *Lechenaultia biloba*, Blue Leschenaultia **05515254**

L. *bi-* = two + *lobus* = lobe

This well-known species generally has a more northern distribution, but is found in a few areas in the north of our area. The colour of its flowers varies hugely from deep, deep blue, often found on gravelly clay, through to white. The paler forms are often found on sandy soils in drier areas. Two colour forms are shown here. It is a perennial herb or slightly woody shrub to about 1 m high and is widely distributed from Geraldton to Esperance and inland almost to Kalgoorlie.

Flowers: Sept.–Dec.

Dampiera

After William Dampier (1652–1715), English navigator, who sailed the north-west coast in 1688 and 1699, collecting some 17 plants in WA. They were the first Australian plants to reach European Herbaria. All the Dampieras in our area have lovely, brilliant blue flowers.

5 *Dampiera alata*, Winged-stem Dampiera **05316464**

L. *alatus* + winged

This is a deep-blue, erect perennial herb growing to about 60 cm, found on clay, sandy or gravelly soil on laterite slopes or flat plains, often in damp locations. It has a long flowering period and has a large range from the northern Darling Range to Albany.

Flowers: Aug.–Dec.

6 *Dampiera hederacea*, Karri Dampiera, Ivy-leaved Dampiera **05474558**

L. *hedera* = ivy, + *-aceous* = like

A vigorous, spreading or climbing perennial herb up to 1 m high, this species is common on sandy, clay or gravel soils, in damp situations, such as swamp edges and creek lines, but it may also be found on roadsides and hill-slopes. It is particularly associated further south with the Karri forest, but here is widespread in other types of forest, woodland and sedgelands. It is found between Pinjarra and Albany.

Flowers: Sept.–Dec.

7 *Dampiera linearis*, Wedge-leaved Dampiera, Common Dampiera **05316405**

L. *linearis* = linear

Despite its common name, the leaves of this, our most common *Dampiera* species, are variable and often far from linear. It is an erect, perennial herb, up to 60 cm. It grows on sandy or gravelly soils, on laterite plains, stony ridges and occasionally swamp edges. Found from Kalbarri to Ravensthorpe.

Flowers: Aug.–Dec.

8 *Dampiera trigona*, Angle-stemmed Dampiera **05515491**

G. *trigonos* = three-angled, triangular

This is a small, slender, tangled perennial herb, to 50 cm, with deep blue-purple flowers, growing on sandy or clay soils, in winter-wet flats and along roadsides. It is found between New Norcia and Albany.

Flowers: Sept.–Jan.

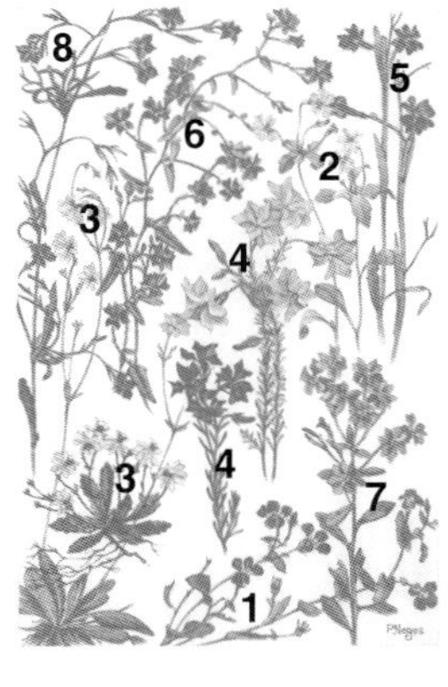

P.Negus

Scaevola

This name, from L. *scaevus* = left-handed, originates from a Roman, C. Mucius Scaevola, who was involved in a plot to assassinate Porsena, who was besieging Rome. He was apprehended and threatened with death by fire if he did not reveal the plot. In reply, he burnt off his right hand. The name refers to the one-sided, fan-shaped corolla.

1 *Scaevola calliptera* **05335884**

G. *calli-* = beautiful + *pteron* = wing (the winged corolla)

This small, creeping perennial is one of the Jarrah forest's prettiest ground-covers. It is only about 30 cm high, but spreads over wide areas amongst the leaf litter on sandy or gravelly soils. It is easily confused with the very similar *S. striata,* royal robe. Common between Perth and Albany.

Flowers: Sept.–Jan.

2 *Scaevola crassifolia*, Thick-leaved Fan-flower **05594995**

L. *crassus* = thick + *folium* = leaf

This is one of the common fan-flowers of the coast, colonising the sand-dunes and limestone cliffs immediately behind the foreshore. It is very similar and easily confused with *S. nitida* (see below), but has broader leaves, smaller flowers that appear later in the year and it tends to grow nearer to the shore. It is dense and usually wind-pruned to a low bush, but can grow to 1.5 m. It is found along the coast, all around the State.

Flowers: Nov.–Feb.

3 *Scaevola glandulifera*, Viscid Hand-flower **05335876**

L. *glandulae* = glands + *fero* = to bear

This sturdy perennial, or slightly woody shrub, has magnificent blue-purple flowers borne on multiple stems. Its hairy stems and leaves are quite sticky. It generally grows to about 50 cm and appears in late spring on sandy clay, or gravelly soils in Jarrah forest and woodlands. It is often quite noticeable along cleared road verges. It is found between Geraldton and Albany.

Flowers: Nov.–Dec.

4 *Scaevola microphylla*, Small-leaved Scaevola **05335922**

G. *micros* = small + *phyllon* = leaf

Another multi-stemmed perennial herb, sometimes with a woody base, with soft, densely hairy leaves and pale blue or mauve flowers. It grows to about 50 cm and is found on sandy or clay soils in Jarrah forest and sometimes along creek-lines. It is found mostly between the Capes and Albany.

Flowers: Oct.–Dec.

5 *Scaevola nitida*, Shining Fan-flower **05515564**

L. *nitidus* = shining, bright; the leaves have a smooth, polished surface

This is the most common species of the coastal heath in our area, growing on sandy soils over limestone. Here, it is usually a dense shrub, wind-pruned to 1 m or less. It is also found in the adjacent Karri forest, however, where it can grow to 3 m when sheltered from the wind. It differs from *S. crassifolia* in having more elliptical leaves that are usually not so quite so thick and stiff. Its flowers are larger and appear earlier in the year. It is found in near-coastal locations from Jurien Bay to Albany.

Flowers: Sept.–Dec.

Diaspasis

G. *diaspasis* = tearing asunder, separating; refers to the deeply divided corolla lobes

6 *Diaspasis filifolia*, Thread-leaved Diaspasis **05797799**

L. *filum* = thread + *folium* = leaf

There is only the one species in this genus. It is a slender, sprawling, perennial herb to 50 cm, with white, pink or purple flowers. It grows on clayey or sandy soils in bogs or winter-wet swamps along the south coast, mainly between Capel and Albany.

Flowers: Dec.–Feb.

6 *Diaspasis filifolia*, Thread-leaved Diaspasis

P.Negus

Stylidaceae

This is a predominantly Australian family, with two genera found in Western Australia, mostly in the south-west. In both *Levenhookia* and *Stylidium* the male and female flower parts have become specialised in a similar way to the orchids, to aid the process of pollination. Most species are found in sandy heathland, or around the damp edges of granite outcrops.

Levenhookia, Styleworts

After Antonius van Leeuwenhoek (1632–1723), Dutch zoologist and microbiologist, discoverer of the capillary circulation of the blood. Styleworts are mostly tiny plants that usually grow in colonies and generally go unnoticed amongst the rest of our more spectacular flora.

1 *Levenhookia pusilla*, Midget Stylewort **05750318**

L. *pusillus* = very small, insignificant

A small, ephemeral annual herb with pink flowers, only 10 cm high, found on sandy or gravelly, lateritic soils over granite. It is common in Jarrah forest or woodland between Geraldton and Esperance.

Flowers: Oct.–Dec.

2 *Levenhookia stipitata*, Common Stylewort **05626757**

L. *stipes* = trunk of a tree; *bot.* stalked

Even tinier than the last, this minute plant is found, usually in large numbers, on gravelly soils in wet depressions or on granite outcrops. Its delicate pink flowers sometimes form multiple flower heads, but are usually unbranched. Widely distributed throughout the south-west.

Flowers: Oct.–Dec.

Stylidium, Trigger Plants

L. *stylus* = style + dim. *-idium,* 2 staminal filaments joined with the style in a column

There are over 200 currently recognised species of these fascinating plants in the south-west, most of which are herbs or climbers. One of the petals forms a tiny labellum that acts as a landing platform for visiting insects. The resulting pressure stimulates the trigger mechanism and the style springs over, dabbing the insect with pollen that it then transfers to another plant on its quest for nectar. Once 'sprung', the style takes some time to re-set itself. The style 'springs' from either the top, bottom or side of the flower, as shown in the diagrams. This characteristic, unique to each species, aids identification.

3 *Stylidium amoenum*, Lovely Trigger Plant **05335949**

L. *amoenus* = charming, showy

Quite a robust, rosetted perennial herb to about 50 cm, with long, slender spikes of mauve flowers. It grows on sand in Jarrah forest and woodland from New Norcia to Albany.

Flowers: Oct.–Dec.

4 *Stylidium calcaratum*, Book Trigger Plant **05750245**

L. *calcar* = spur, spurred

A delicate annual to about 20 cm with a spur on the corolla. Flowers pink or white, often covering large areas around ephemeral pools. Found in damp places and around granite outcrops throughout the south-west.

Flowers: Oct.–Jan.

5 *Stylidium piliferum*, Common Butterfly Trigger Plant **05515327**

L. *pilus* = hair + *fero* = to bear

This species is covered in dark hairs. It grows to about 50 cm and is found on sandy or lateritic soils in Jarrah or Peppermint woodland. It is very common throughout the south-west.

Flowers: Sept.–Nov.

6 *Stylidium petiolare*, Horn Trigger Plant **05771951**

L. *petiolus* = little foot; *bot.* petiole = leaf stalk

Another small perennial, only 15 cm high, but growing in attractive colonies in swampy areas and on granite outcrops. Found widely dispersed through the south-west.

Flowers: Aug.–Oct.

7 *Stylidium schoenoides*, Cow Kicks **05750253**

Genus *Schoenus* + *-oides* = like; refers to the long, narrow leaves

This is one of our more spectacular trigger plants, with a much bigger flower than most, even though it is low to the ground, only about 20 cm high. It likes sandy soil in open heath or woodland and occurs between Dongara and Esperance. The large flower of this species is ideal for examining the intriguing trigger mechanism (see illustration 1, 2 & 3).

Flowers: Oct.–Dec.

8 *Stylidium spathulatum* subsp. *spathulatum* **05515351**
Creamy Trigger Plant

L. *spathula* = spoon; the leaves are spoon-shaped

A lovely rosetted trigger plant up to 40 cm tall that grows in colonies on loamy soils in Jarrah forest, or on granite outcrops. The flowers can be white or cream, but are generally yellow here. It is found between Bunbury and Albany.

Flowers: Oct.–Nov.

1 *Levenhookia pusilla*
Midget Stylewort

2 *Levenhookia stipitata*
Common Stylewort

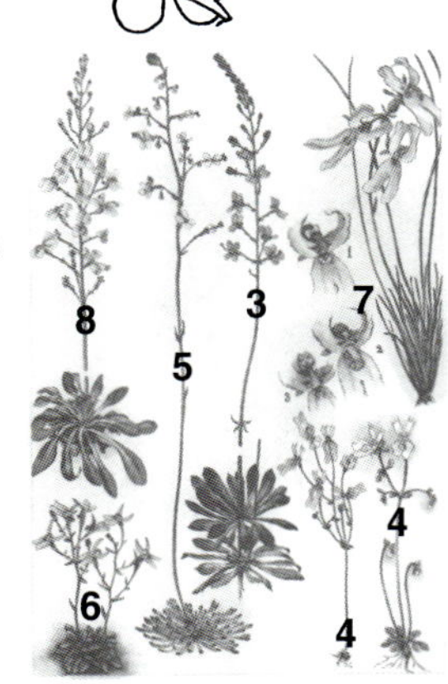

1
2
3

1 *Stylidium caespitosum*, Fly-away Trigger Plant  **05515483**

L. *caespes* = mound, thatch of turf; *bot.* caespitose = tufted

A small, rosetted perennial herb to 30 cm, growing on sandy clay soils in winter-wet depressions or seepage areas. It is found between Bunbury and Albany.

Flowers: Nov.–Jan.

2 *Stylidium fasciculatum*, Pale Beaked Trigger Plant 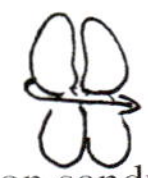 **05515335**

L. *fasciculus* = small bundle; refers to the leaf whorls up the stem

An erect little herb to 60 cm, with bright pink flowers, growing on sandy soils along limestone cliffs and hill-slopes. It grows from the Capes coast to Albany. There are several species in the 'Beaked Trigger Plant' group that are all superficially similar. Their names may change as a result of ongoing taxonomic revision within the genus.

Flowers: Oct.–Dec.

3 *Stylidium junceum* subsp. ***junceum***, Reed Trigger Plant 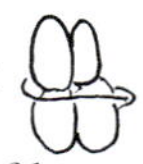 **05515416**

L. *junceus* = rush-like

This is a tall species up to 1 m high with a basal rosette of leaves and long stems ending in broad, terminal flower spikes. These are usually pink, but may be white, cream, yellow, or mauve. Found on lateritic or sandy soils over limestone, in heath or swampy areas. It is common from Jurien Bay to Albany.

Flowers: Oct.–Dec.

4 *Stylidium laciniatum*, Tattered Trigger Plant 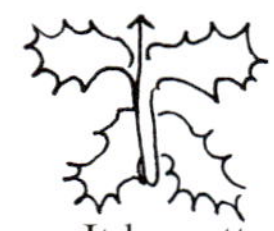 **05917530**

L. *lacinia* = flap of a garment; *bot.* used for an edge that appears torn

This is a robust, climbing perennial of peaty sands and swamp edges. It has attractive pink, frayed-edge flowers and can reach 3 m in height. It is a plant of south coast swamps found between Forest Grove and Albany.

Flowers: Dec.–Feb.

5 *Stylidium violaceum*, Violet Trigger Plant  **05577276**

L. *violaceus* = violet

This attractive species grows to about 50 cm with widely spaced flowers that bloom a few at a time and a basal rosette of bluish-green leaves. It is found on sandy soils in woodland and heath from Bunbury to Albany.

Flowers: Oct.–Dec.

6 *Stylidium crassifolium*, Thick-leaved Trigger Plant 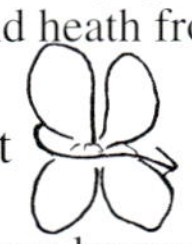 **05624991**

L. *crassus* = thick + *folium* = leaf

This is a stout perennial with a rosette of broad, green leaves and a tall flower spike up to 1 m tall. It favours damp locations, growing on loam or sandy clay in winter-wet seepage areas and shady slopes. Found between Perth and Albany and along the south coast to Esperance.

Flowers: Aug.–Oct.

7 *Stylidium dichotomum*, Pins-and-needles  **05626765**

G. *dicha* = in two ways + *tomios* = cut up; forked or branching

This is a creeping, cushion-like perennial with narrow leaves that become hard and needle-like as they dry in summer. Its small flowers vary in colour throughout its range from white, through pink, yellow and orange. The flowers have noticeably long, hairy calyx tubes. It grows on sand or clay soils over laterite or granite, also in swampy areas and is found between Eneabba and Esperance.

Flowers: Nov. – Dec.

8 *Stylidium eriopodum*  **05626803**

G. *erion* = wool + *pous, podos* = foot

The oddly shaped petals of this trigger plant are like those of the aptly-named Boomerang Trigger Plant, *S. breviscapum*. Like this species, it also has tough, wiry, above-ground stilt roots. It is a perennial and the roots elevate the plant above the soil, helping to keep it cool through summer. It is small, growing only to about 20 cm, but its reddish stems and whorls of leaves are distinctive. It grows on sandy, gravelly soil on and around granite outcrops and stony slopes. It mainly occurs inland between Dalwallinu and Jerramungup, but has also been collected in the Perth hills and down the Capes Coast.

Flowers: Nov.–Dec.

8 *Stylidium eriopodum*

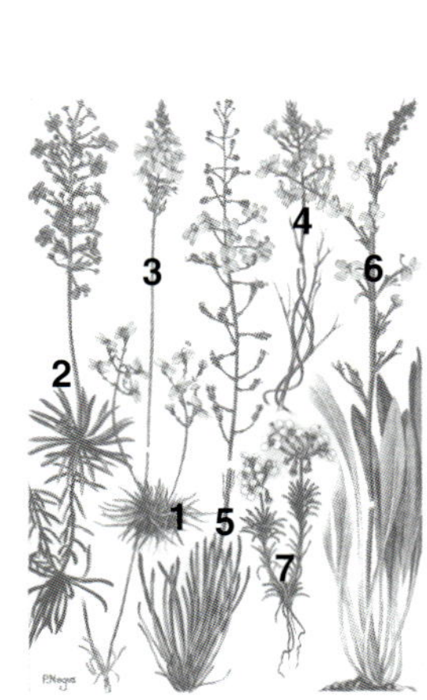

P.Negus

Asteraceae, Daisies

G. *aster* = a star

This is one of the world's largest and best known plant families, with up to 20,000 species spread over every continent. Western Australia is famous for its fields of everlastings that may form acres of massed colour in the outback in spring. In the south-west corner we do have a few everlastings that form large aggregations, but most of our daisies occur individually or in small groups.

The flower heads of members of the daisy family actually comprise many individual flowers, all sitting on a fleshy receptacle and surrounded by whorls of involucral bracts.

Trichocline

G. *thrix, thricho* = hair + *cline* = couch or bed; the involucre is covered with white wool

1 *Trichocline spathulata*, Native Gerbera **05335906**

spathulatus = spoon-shaped

This perennial herb is usually found singly on loamy soils in Jarrah forest, though it may appear 'en masse' after a fire. The basal rosette of leaves is evident throughout the spring, but they are beginning to wither by the time the flowers appear in late spring and summer. The flowers are tall and prominent, reaching up to 1 m high. It grows between Jurien Bay and Albany.

Flowers: Nov.–Jan.

Olearia

L. *olea* = olive tree; the first named species has leaves resembling those of the olive

2 *Olearia axillaris*, Coastal Daisy Bush **05611911**

L. *axilla* = arm-pit; *bot.* axillary–growing from the axil, the angle between the leaf stalk and the branch or stem.

Also known as Native Rosemary, because of the shape of the leaves and their powerful aroma, this is one of the dominant species on the coastal limestone. It has lovely, grey-green foliage and usually grows as a compact, rounded shrub to about 2 m on the coastal heath, though it can grow taller and more straggly in forest shade. The yellow-brown flower heads are small, but numerous. It is found all around the south-west coast.

Flowers: April–June

3 *Olearia elaeophila* **05594928**

Formerly *O. heleophilus:* G. *heleos* = marsh + *phileo* = to love

A slender, branching shrub to 1 m or so, bearing small, very pale blue flowers in late summer and early autumn. It grows on sandy or clay soil, often in wet areas and swamps, hence the name, but is also found in drier country and rocky outcrops from Jurien Bay to Albany.

Flowers: Feb.–April

4 *Olearia paucidentata*, Autumn Scrub Daisy **05577489**

L. *paucus* = few, little + *dens, dentis* = tooth (on the leaves)

Very similar to the last species, but this one is more robust, with larger flowers. These appear in late spring and early summer in our area and its common name would be more applicable to *O. elaeophila.* It grows to over 1 m, again often in damp situations, or in shady forest, on a variety of soils and is again found between Jurien Bay and Albany.

Flowers: Nov.–Feb.

5 *Olearia rudis*, Rough Daisy Bush, Azure Daisy Bush **05577373**

L. *rudis* = rough, wild

This is a magnificent plant, its second, more complimentary common name suiting it much better. Its large, bright purple flowers with their yellow centres are very distinctive and are conspicuous amongst the understorey foliage of the Karri forest, where it often grows to 1.5 m high on the calcareous loamy soil. Its large leaves are quite aromatic and its lower stems woody. Found from Dongara to Ravensthorpe.

Flowers: Oct.–Dec.

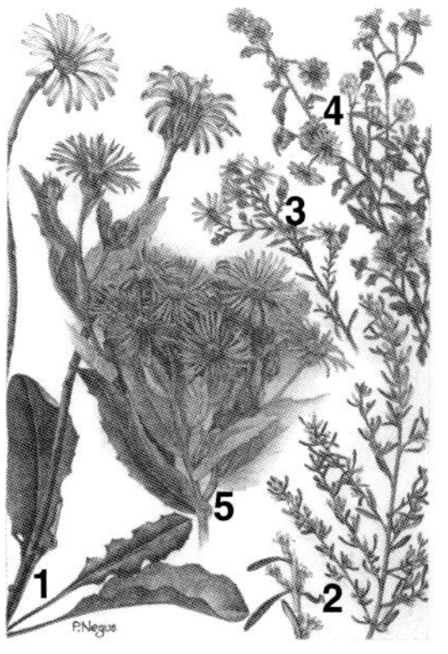

P.Negus

Asteridea

L. *asteridea* = star-like

1 *Asteridea pulverulenta*, Common Bristle Daisy

L. *pulverulentus* = full of dust, powdery

This is a sturdy annual, standing up to about 40 cm high. It is frequently found on sandy soils over limestone or on granite, coastal dunes and sandplains and occurs between Eneabba and Albany.

Flowers: Oct.–Dec.

Brachyscome

G. *brachys* = short + *come* = hair; refers to the short ring of hairs (the pappus) around the seed (achene)

2 *Brachyscome iberidifolia*, Swan River Daisy **05515270**

From the genus: *Iberis* + L. *folium* = leaf

This common and attractive little annual herb adorns the floor of heathland, woodlands and open forest all through spring. It grows to about 30 cm on sandy or clay soils and is widely distributed throughout the south-west.

Flowers: Sept.–Nov.

Rhodanthe

3 *Rhodanthe citrina* **05625246**

Small, erect annual herb, with brilliant yellow, papery flowers, to 60 cm. It is common along roadsides and in sandy country near the coast. Widespread throughout the south-west.

Flowers: Sept.–Nov.

Leucophyta

G. *leucos* = white, grey + *phyton* = plant

4 *Leucophyta brownii*, Cushion Bush **05456851**

After Robert Brown (1773–1858), naturalist on Matthew Flinders' expedition and the most distinguished early collector of WA flora

This silvery-grey, tightly compact coastal shrub is attractive and distinctive, often seen in cultivation as a border plant. The individual yellow flowers are tiny. It is usually wind-pruned to a low, compact bush and is found on dunes and limestone sands all around the coast from Perth to Esperance.

Flowers: Nov.–Feb.

Siloxerus

G. *stylos* = a pillar (*bot.* style) + *ogceros* = swollen; swollen base of the style

5 *Siloxerus humifusus*, Procumbent Siloxerus

L. *humi-* = on the ground + *fusus* = spread out; *bot.* having a low, spreading growth habit

This is a common small herb, forming dense mats especially on granite outcrops and in moist situations on sandy soils from Jurien Bay to Esperance. The flowers are white or cream. Another similar species, *S. filifolius*, which has rounded, more succulent leaves, is also found in the area.

Flowers: Oct.–Dec.

Senecio

L. *senex* = old man; refers to the white pappus of *Senecio vulgaris,* Common Groundsel (originally known as Erigeron)

6 *Senecio lautus* subsp. *maritimus*, Coastal Groundsel **05515548**

L. *lautus* = elegant, splendid; *maritimus* = of the sea

A prominent, erect or sprawling annual or perennial herb, around 60 cm high, growing on limestone and coastal dunes. The leaves and habit are very variable, depending on conditions. It grows low with fleshy leaves when exposed to the salt and wind, but tall and slender in more sheltered locations. The subspecies is found all around the coast from Shark Bay to the Great Australian Bight.

Flowers: Sept.–Jan.

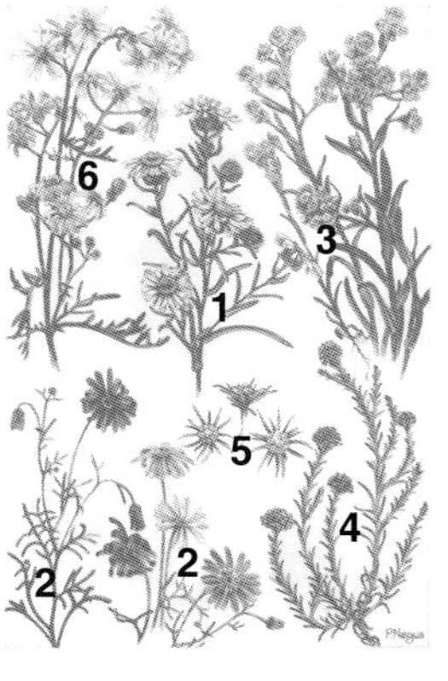

P.Negus

1 *Senecio ramossissimus*, Auricled Groundsel **05577446**

L. *ramosus* = much branched (superl. form, very much branched)

This is a very large and impressive perennial herb, to 2 m or more, growing here in the Karri forest on sandy limestone soils. It is especially rampant after fire. It is found in near coastal locations from Yanchep to Hopetoun.

Flowers: Nov.–Dec.

Hyalosperma

G. *hyalos* = glass + *sperma* = seed; refers to the transparent wing of the achene (seed)

2 *Hyalosperma cotula* **05577292**

L. *cotula* = small vessel or cup

A small everlasting to 25 cm, growing in large numbers in sandy soils between Kalbarri and Albany.

Flowers: Oct.–Dec.

Podolepis

G. *pous, podos* = foot + *lepis* = scale; the inner involucral bracts are stalked

3 *Podolepis gracilis*, Slender Podolepis **05577241**

L. *gracilis* = thin, slender

This is a rather lovely slender, erect annual herb, with white or pink flowers, to about 50 cm. It grows on sandy soils in many situations, here mainly on the coastal limestone ridge where it mingles with lobelias, trigger plants and other daisies to form a colourful floral carpet in late spring. It also grows on granite outcrops and is common between Jurien Bay and Albany.

Flowers: Nov.–Jan.

4 *Podolepis lessonii* **05625238**

After Rene Primevere Lesson (1794–1849), pharmacist, ornithologist, botanist; sailed on expeditions of Duperrey (1822–25) and D'Urville (1826–29)

This is a dainty little herb up to 40 cm high that grows in quantity along the bush tracks in open woodland on a variety of soils. Widespread throughout the south-west.

Flowers: Aug.–Nov.

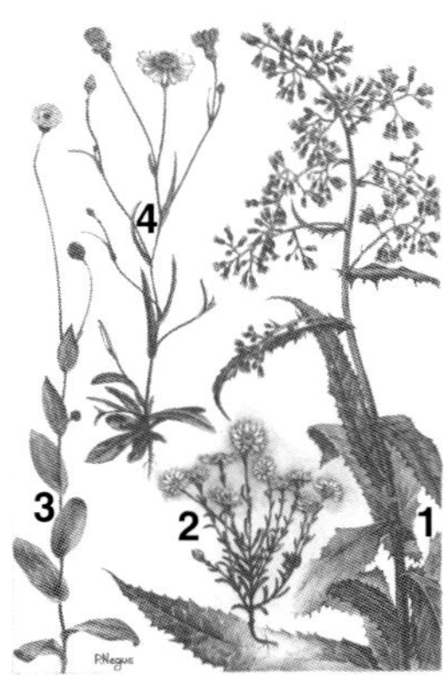

P.Negus

Craspedia

G. *craspedon* = border, edge; refers to the fringed leaves of the original species, or to the feathery pappus

1 *Craspedia variabilis*, Billy Buttons **05750288**

L. *variabilis* = variable

Billy buttons is one of our prominent early-flowering daisies, with tightly packed flower heads standing out above lower vegetation. The flower colour varies from pale lemon through to deep golden yellow. It grows to about 60 cm, on sandy or lateritic soils, in Jarrah forest or open woodland, but also on wet sand flats. It is found from Bindoon to Albany.

Flowers: Aug.–Nov.

Ozothamnus

2 *Ozothamnus cordatus* **05594960**

L. *cordis* = heart; heart-shaped

A tall, ascending or sprawling, slender shrub, with attractive, silver-grey foliage and stems up to 1.5 m high. It grows in loose, tangled clumps on sandy limestone soils, in forest or on coastal heath. It has a near coastal distribution from Jurien Bay to Albany.

Flowers: Oct.–Jan.

3 *Ozothamnus ramosus* **05577470**

L. *ramosus* = much branched

Another lanky species, this is similar to the last, growing to about the same height. The flowers are pinkish white, instead of yellow and white and its leaves are much narrower. It grows on a variety of soils, mostly between Bunbury and Albany. Both these plants make most attractive additions to floral arrangements.

Flowers: Nov.–Feb.

Lagenophora

G. *lagenos* = flagon + *phoreo* = to bear; refers to the shape of the achenes (seeds)

4 *Lagenophora huegelii* **05750326**

After Carl Alexander Anselm, Baron von Hügel (1796–1870), Austrian traveller and naturalist, visited the Swan River and King George Sound in 1833

This rosetted little daisy looks as if it should be a weed, but is actually a common local inhabitant. It has white, pink or purplish flowers that grow erect to about 30 cm. It is found on a variety of soils, in woodland and heath and is widespread throughout the south-west from Geraldton to Bremer Bay.

Flowers: Aug.–Nov.

Cotula

L. *cotula* = small vessel or cup; refers to the flower shape, or maybe to the stem-clasping leaf bases

5 **Cotula turbinata*, Funnel Weed **05623979**

L. *turbo, -inis* = a spinning top; shaped like a top

This is an annual herb that grows to 40 cm but is often smaller. It is common on lawns and along roadsides in the lower south-west between Perth and Albany. As with so many of our weeds it is native to South Africa.

Flowers: July–Oct.

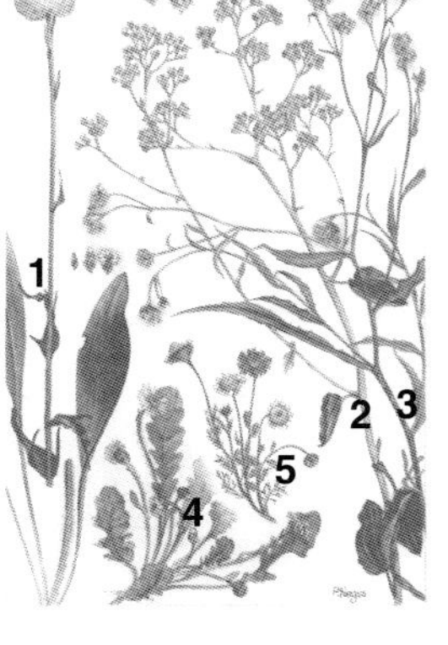

P.Negus

Weeds

A very large number of alien (weed) species grow in the area, only a few of which are illustrated in this volume. A few have been incorporated in the main part of the book, but most are treated separately here. All alien species throughout the book are marked with an asterisk (*). The arrangement is rather 'ad-hoc', determined – we admit – by the order in which we decided to include them, rather than by taxonomic logic. Members of the same family, however, are grouped together. The book *Western Weeds* by B.M.J. Hussey et. al. was the source of much of the information in this section.

Our focus with the weeds has been to show the more obvious plants that are seen in, or on the edge of bushland habitats, and especially those that may be mistaken for indigenous species. Some of them are benign, but others pose serious threats to the integrity of our natural bushland. Recognition of these weeds is the first step towards their control.

DP = Declared Plant. Land holders legally obliged to control at own expense.

PP = Pest Plant. Shire council authorised to enforce control within its boundaries.

Monocotyledons

Amaryllidaceae, Daffodil family

1 *Amaryllis belladonna*, Easter Lily

This colourful lily appears, often as a complete surprise, in early autumn, along roadsides and in fields, betraying the former presence of old settlements long after any evidence of their structure has disappeared. It only occasionally spreads into the bush. At the time of flowering it is leafless, its leaves appearing later in the year. Native to South Africa.

Flowers: Mar.–May

Araceae, Arum family

2 *Zantedeschia aethiopica*, Arum Lily ***DP*** **05625084**

One of our most devastating weeds. This species spreads vegetatively by means of underground corms that choke out native plants and also by seeds that are dispersed by birds. It is found throughout the area in damp habitats, but is also spreading increasingly onto the drier sandy soils of the Leeuwin–Naturaliste Ridge where it now frequently appears amongst dense native bushland. Native to South Africa.

Flowers: Aug.–Nov.

Asphodelaceae, Asphodel family

3 *Trachyandra divaricata*, Dune Onion Weed **05799163**

This is a widely branching perennial, growing to about 1 m, with flat, basal leaves that can smother other vegetation. It is common on calcareous, sandy soils, along the coast in heathland between the dunes. Native to South Africa.

Flowers: Year round, but mainly in spring

Iridaceae, Iris family

There are many weed species from this family prevalent in the south-west, mostly originating from South Africa. Separating them can be confusing. Ironically, several are now quite rare in their homeland.

4 *Gladiolus undulatus*, Wavy Gladiolus

Common along roadsides and in wetlands and starting to invade adjacent bushland. it is usually a little less than 1 m high. Native to South Africa.

Flowers: Oct.–Dec.

5 *Moraea (Homeria) flaccida*, Cape Tulip ***DP*** **05624975**

A weed of pasture, woodlands, granite rocks and limestone heath. This species is not widely established here, but has the potential to become a serious pest in the bush, so keep a lookout for it and remove speedily if you find it! It is a slender plant, up to 70 cm tall. Native to South Africa.

Flowers: Sept.–Nov.

5 *Moraea (Homeria) flaccida*, Cape Tulip

P.Negus

1 ***Ferraria crispa*****,** Black Flag

The succulent foliage and distinctive flower of this plant make it easy to recognise. It is fairly small, growing to about 60 cm. Found here, so far mainly on roadsides and wasteland. Native to South Africa.
Flowers: Sept.–Oct.

2 ***Sparaxis bulbifera*****,** Freesia **05625130**

This small, attractive garden flower, about 30 cm tall, has become quite a serious weed along roadsides, in fields and adjacent bushland, spreading from corms and seeds. It is native to South Africa.
Flowers: Aug.–Nov.

3 ***Romulea rosea*****,** Guildford Grass

This is a most familiar weed, appearing on most peoples' lawn at some stage! It is also found in pastures and woodlands and limestone heath. The small flowers are at ground level, but have the interesting habit of elongating up to 20 cm or so with maturity and bending over, pushing the seed back under the surrounding vegetation. Native to South Africa.
Flowers: Aug.–Oct.

4 ***Ixia*** **aff.** ***maculata*****,** Variable Ixia **05625319**

Three or four species of Ixia have become naturalised in WA, including *I. maculata*, as well as hybrids involving *I. maculata*. This one blooms on slender, tremulous stems and has a spike of several nodding flowers in shades of white, pink, purple or yellow. Up to 1 m tall. Native to South Africa.
Flowers: Oct.–Dec.

5 ***Chasmanthe floribunda*****,** African Cornflag

Similar to Watsonias (below), but with narrower flowers that appear earlier in the year. This is a tall species with flower spikes to 1.5 m. and occurs in urban bushland, road verges and wasteland. Mainly orange, but occasionally yellow.
Flowers: Aug.–Oct.

Watsonia, all species **PP**

There are several naturalised species of *Watsonia* in the south-west, all garden escapes and often hybridised. Like many of our aliens they are native to South Africa. All are serious weeds of roadsides and adjacent bushland.

6 ***Watsonia meriana*** **var.** ***bulbillifera*****,** Bulbil Watsonia **05624797**

This is the common orange Watsonia, not to be confused with the earlier-flowering cornflag that belongs to a different genus (see above). The flower spikes grow to 1 m and many small corms form at each node up the flower stalk to ensure that it spreads very efficiently!
Flowers: Oct.–Dec.

7 ***Watsonia versfeldii*** **05624800**

This is another common species along our roadsides, with tall pink or white flower spikes growing to 1.5 m. The very similar *W. borbonica* is also present in the area.
Flowers: Oct.–Nov.

Asparagaceae, Asparagus family

8 ***Asparagus asparagoides*****,** Bridal Creeper

Bridal Creeper ranks with Arum Lily and Watsonia as one of our area's worst weeds! It creeps and climbs its way over native vegetation in many places, especially old settlement sites, smothering plants and preventing the growth of new seedlings. Its seeds are dispersed in the droppings of birds, so it can easily be spread into undisturbed bushland. Another native of South Africa.
Flowers: Aug.–Sept.

6 ***Watsonia meriana*** var. ***bulbillifera***, Bulbil Watsonia

P. Negus

Dicotyledons

Onagraceae, Willow-herb family

This family includes several species of Evening Primrose as well as the European willow-herbs, some of which have become naturalised in Australia. There are also several native species of willow-herb (*Epilobium* spp) in WA.

1 *Oenothera glazioviana*, Tall Evening Primrose **05795176**

This plant produces leafy flowering stalks to 1.6 m and is common along roadsides and wasteland between Perth and Albany. Most *Oenothera* species here originate from America, but this one is probably from Europe.

Flowers: Nov.–March (mostly)

Oxalidaceae, Wood-sorrel family

About fifteen species of *Oxalis* (Wood-sorrel) are currently known in the south-west and have yellow, pink, purple or white flowers. Most are garden escapes and most are native to South Africa.

Shown here are:

2 *Oxalis pes-caprae*, Soursob ***DP*** **05625092**

3 *Oxalis purpurea*, Largeflower Wood-sorrel, or Four O'clock

Both flower in early spring.

Apocynaceae, Periwinkle family

4*Vinca major, Blue Periwinkle **05624908**

This is a garden escape and a common weed near settlement sites. It is a prostrate perennial with arched stems that take root when in contact with the ground. The flower stems are about 50 cm high. It spreads over wide areas in shady sites. Native to the Mediterranean.

Flowers: Aug.–Nov.

Primulaceae, Primrose family

5*Anagallis arvensis, Pimpernel, Red and Blue

Plants of each flower colour are varieties of the same species, the blue variety often being the more robust plant. They are usually only about 20 cm high and are common in bushland, but not harmful.

Flowers: Aug.–Jan.

The following are examples of Australian plants that have become established outside their normal ranges and thus have entered the realm of weeds.

Myrtaceae, Myrtle family

6 *Leptospermum laevigatum*, Victorian Teatree **05624096**

This large shrub, 3 or 4 m tall and 2 m wide, was introduced to WA as a garden plant, being particularly useful for hedgerows, but has escaped to become a major bushland weed. It is found around old settlements, but is very invasive and is spreading rapidly along roadsides and into heathland and woodlands between Jurien Bay and Albany.

Flowers: Aug.–Oct.

7 *Leptospermum erubescens*, Roadside Teatree **05626986**

This is an interesting case of a shrub that is native to other parts of Western Australia, but which here appears to be outside its natural range. It has colonised a sand quarry on Bussell Highway south of Forest Grove in large numbers and appears to be spreading, though at present only in the disturbed area. It may be present elsewhere in the area and we await more reports of its presence that may indicate that it belongs here after all.

Flowers: July–Sept.

Pittosporaceae, Pittosporum family

8 *Pittosporum undulatum*, Sweet Pittosporum, Mock Orange **05624924**

This garden escape is a bushy tree up to 5 m tall with terminal clusters of sweet-smelling creamy flowers that develop into orange fruits. It is gaining a foothold in moist locations where it grows vigorously, especially along the Margaret River and at Augusta. It is capable of spreading rapidly and is potentially a very serious weed. In other areas it has become completely dominant, its broad leaves ensuring that other plants are shaded out. It is native to south-eastern Australia.

Flowers: Oct.–Nov.

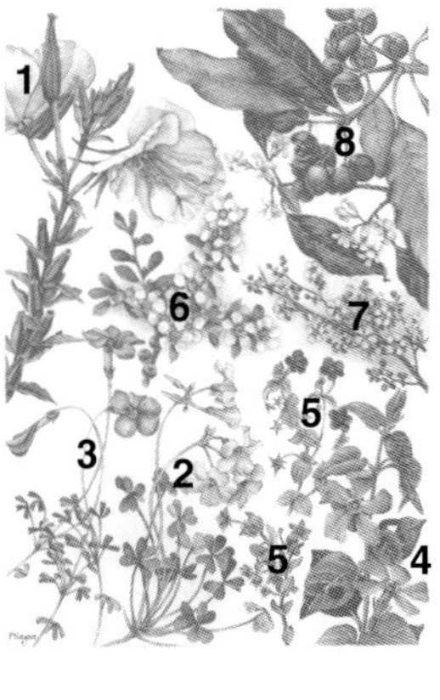

P.Negus

Papilionaceae, Pea family

This is one of the world's largest plant families, so it is not surprising that many of our weeds belong to it. Many field weeds such as medics and clovers belong to this group, but here we show some of the species more inclined to colonise the bushland.

1 *Dipogon lignosus, Dolichos Pea **05626641**

This is a rampant climbing perennial that smothers the vegetation on which it finds support. It may be white, pink or purple and was introduced from South Africa as a garden plant.

It is most common in the wetter Karri forest.

Flowers: Sept.–Nov.

2 *Genista monspessulana, Broom **05624894**

This is one of several species of Broom that are naturalised and occur along roadsides and on wasteland in the wetter parts of the south-west. It is an open, spreading shrub to 3 m high and will form dense thickets on open ground if not discouraged. Native to the Mediterranean.

Flowers: Sept.–Dec.

3 *Podalyria sericea **05611989**

This is a rather lovely silvery-grey shrub to about 1.5 m. It appears to grow in only a few places along Caves Road and on the corner of Boodjidup and Gnarawary Roads, but is included here as it is a very striking plant. It is easily mistaken for a native. It is, however, a garden escape that is native to South Africa.

Flowers: July–Sept.

4 *Chamaecytisus palmensis, Tree Lucerne, Tagasaste **05624789**

Grown extensively as a fodder crop, Tagasaste seeds prolifically and has spread along roadsides and into disturbed bushland. It is a tall shrub or small tree up to 6 m, with open, drooping branches. Native to the Canary Islands.

Flowers: July–Oct.

Gentianaceae, Gentian family

5 *Centaurium tenuiflorum, Slender Centaury **05795109**

This is the most common of several introduced species of centaury. There are also two native species in WA. It is an annual herb, about 50 cm tall and grows along roadsides and in other sites. Native to Europe.

Flowers: Oct.–Jan.

Scrophulariaceae, Snapdragon family

6 *Bartsia trixago, White Bartsia **05771846**

This annual herb, about 30 cm high, is quite common on the sandy coastal heath of the Leeuwin–Naturaliste Ridge and is easily mistaken for a native plant. It is semi-parasitic on the roots of surrounding vegetation. Native to the Mediterranean and the Middle East.

Flowers: Nov.–Jan.

Caryophyllaceae, Carnation family

7 *Petrorhagia dubia, Velvet Pink **05624681**

A slender upright herb, about 50 cm high growing on wasteland and road verges, but often also well into bushland, especially Peppermint woodland. Native to Europe.

Flowers: Sept.–Dec.

Asteraceae, Daisy family

This very large, worldwide plant family is represented here by a considerable number of weed species. The daisies are often 'opportunists' that can quickly colonise a wide variety of habitats and are usually able to thrive in alien environments.

8 *Arctotheca calendula, Capeweed **05624134**

Capeweed is a familiar sight in spring when it carpets the paddocks with its golden flowers. Although mainly a weed of pastures it also colonises waste ground, roadsides and bare ground in bushland. Native to South Africa.

Flowers: Sept.–Oct.

9 *Arctotheca populifolia, Dune Cabbage **05750261**

This tough, ground-hugging plant is so much a part of our coastal vegetation that it is often thought to belong here. Its flowers are a lot smaller than those of Capeweed, but its silvery-grey leaves are attractive. It grows on the fore-dunes between Perth and Esperance and performs a useful role in helping to stabilise them. It is also native to Southern Africa.

Flowers: July–Oct.

10 *Senecio elegans, Purple Groundsel **05750296**

What a pity this magnificent plant is a weed! Its dazzling magenta flowers add beautiful splashes of colour to the coastal dunes in spring. However, not all weeds are harmful and this species does not seem to encroach seriously on our native vegetation, at least in this area. It is quite a stout plant that can grow to 1 m high. Another South African native.

Flowers: Sept.–Mar.

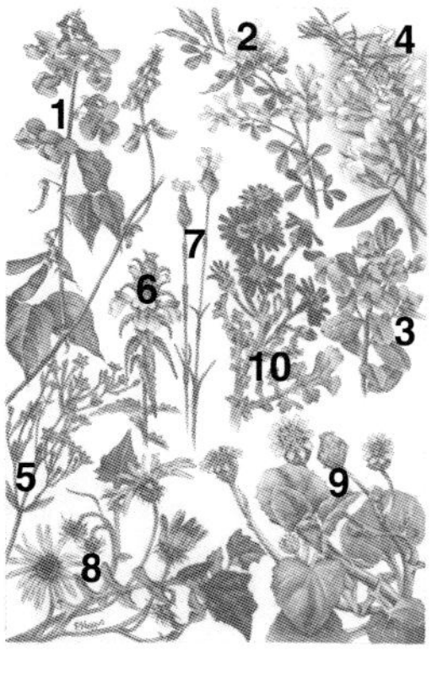

P.Negus

Solanaceae, Potato family

1 **Solanum linneanum*, Apple of Sodom ***DP, PP*** **05625254**

A descriptive name for a very nasty plant! It is a stout and extremely thorny shrub that can grow up to 2 m or more. Every part of this plant seems to bear spines. If you can, however, you should pull up and destroy any that you see. Take particular care to destroy the seeds. It grows along roadsides and into coastal heath and there is not yet very much of it in our shire so containment is still possible. Native to South Africa.

Flowers: Aug.–Oct.

Polygonaceae, Dock family

2 **Emex australis*, Double Gee ***DP, PP*** **05625297**

Though not as big a problem here as in some parts of the state, Double Gee is in the area and should be destroyed wherever found. It is normally a prostrate annual and the example illustrated here is quite small, but it can grow into a stout upright plant of up to 1 m. Vehicles readily transport it as the seed's rigid spines stick easily into tyres and anything else! Native to South Africa, it was originally introduced as a salad vegetable!

Flowers: July–Oct.

8 **Verbascum virgatum*, Twiggy Mullein

Fumariaceae, Fumitory family

Five species of Fumitory have been identified in WA, but the pink species are very difficult to separate.

3 **Fumaria capreolata*, White Fumitory **05625122**

This one is easy to identify as it is the only white species. It is a sprawling climber and reaches 1 m or more in length. A common garden weed that can spread into the bush when garden rubbish is dumped. Native to Eurasia.

Flowers: Aug.–Nov.

4 **Fumaria muralis*, Wall Fumitory **05624886**

The most common of the pink fumitories, this slender annual herb climbs over other plants and up walls and fences to about 1 m high. Native to Europe.

Flowers: Aug.–Dec.

Caryophyllaceae, Carnation family

5 **Silene gallica* var. *gallica*, French Catchfly **05625106**

This is a rather attractive little relative of the campions (see Velvet Pink, previous page). It is a slender annual with slightly sticky (viscid) stems and leaves and grows to about 50 cm. It is found along roadsides and wasteland and spreads into coastal woodland on sandy soils. Native to Europe.

Flowers: Sept.–Nov.

Euphorbiaceae, Spurge family

The spurges (*Euphorbia*) all have a milky sap that can be corrosive, so it is wise not to pick them. Their flowers are tiny and surrounded by more conspicuous floral leaves. WA has about 40 species, of which a quarter is introduced.

6 **Euphorbia paralias*, Sea Spurge **05626633**

Sea Spurge is a clumped, fleshy perennial, with terminal flower heads, that is a common weed of beaches along the south coast. The stems turn red as they age and with their yellow-green flowers can make an attractive sight along the dunes. Native to the Mediterranean.

Flowers: Nov.–Jan.

Boraginaceae, Borage family

7 *Echium plantagineum*, Paterson's Curse, ***DP*** **05626617**

A sturdy annual with bristly leaves and stems, this plant grows to about 60 cm. It has become widespread in paddocks, but also multiplies rapidly along roadsides and other disturbed areas, crowding out native plants, especially the smaller annuals. Luckily, it has not yet taken hold too much in our area. It is easy to identify and should be removed wherever it is found.

Flowers: Oct.–Jan.

Scrophulariaceae, Snapdragon family

8 **Verbascum virgatum*, Twiggy Mullein **05626811**

Mulleins have a basal rosette of leaves and a tall flower spike. This one grows up to 2 m. It is found in disturbed areas, along roadsides and spreading into the bush along firebreaks. Native to Europe.

Flowers: Dec.–Feb.

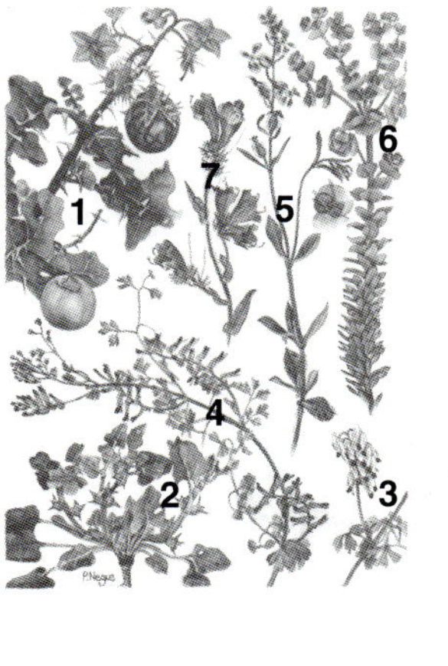

P.Negus

Galls

Things you see in the bush are not always what they seem! What look like rosy little fruits on *Melaleuca incana*, or deep pink flowers gone mad with too many petals on *Boronia crenulata*, are actually structures known as galls. These strange outgrowths take many forms, as can be seen from the examples we have illustrated here. As well as resembling fruits, buds or flowers, they may take the form of prolific leafy outgrowths, bumpy protrusions or warts on leaves, gnarled, woody lumps on stems or branches, or even underground nodules growing on roots, as in the case of the root nodules of legumes (the pea family) that nitrogen-fixing bacteria form.

Galls can be defined as any abnormal growth caused by a plant's reaction to attack by a variety of organisms. The most common of these organisms are many different species of insects, but other parasites such as mites, nematodes, bacteria and fungi may also be responsible. They irritate the plant's tissues causing either the proliferation, or excess growth of its cells. It is this that forms the gall structures. In this way, the plant attempts to contain the gall-producing organism within the boundaries of the gall and so limit its spread.

Galls provide useful shelter for the organisms that cause them and they are sometimes also a source of food. If you cut open a gall you will often find that it is hollow and may still contain one or more insect larvae. Once the grub has grown and pupated within the shelter of the gall the adult insect emerges and eats its way out. The relationship between gall-producing organism and plant seems generally to favour the organism, though there are occasionally advantages to the plant. In the case of the legume's root nodules, the residing bacteria 'fix' atmospheric nitrogen to produce soluble nitrates that the plant can use. This is a more mutually beneficial relationship than the more usual one between parasite and host.

Some of the specimens illustrated here (♦) were sent to the Western Australian Museum and were found to contain a variety of gall-producing insects, including flies (Diptera), wasps (Hymenoptera) and moths (Lepidoptera); a gall-forming mite (on *Corymbia calophylla*); and also several wasps that may be parasites of the gall-producers.

Plants illustrated:

1 *Leucopogon elatior*
2 *Leucopogon revolutus* ♦
3 *Corymbia calophylla* ♦
4 *Acacia myrtifolia*
5 *Hypocalymma robustum*
6 *Acacia extensa*
7 *Bossiaea ornata*
8 *Boronia crenulata*
9 *Melaleuca incana* ♦
10 *Leucopogon capitellatus*
11 *Banksia ilicifolia* ♦
12 *Olearia axillaris* ♦
13 *Acacia cochlearis*
14 *Agonis parviceps*

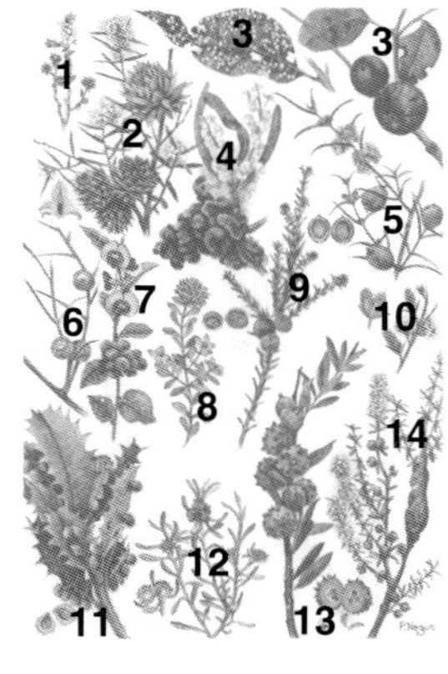

P. Negus

Habitats

A limestone ridge up to 200 m high that runs north to south between Cape Naturaliste and Cape Leeuwin dominates the topography of the western part of the Augusta–Margaret River Shire. To the east of this major feature flat swamplands and sandflats give way to undulating country with low, lateritic hills intersected by gullies and streams. The underlying basement rock throughout much of the area is ancient granite-gneiss and granite outcrops punctuate the landscape in places.

The following photographs and descriptions are a guide to some of the main vegetation types that are found here. There are of course many variations within these, but we hope they will help you to become familiar with the assortment of plants to expect in different situations. The 'Typical species' represent only a tiny fraction of those found in any one habitat type and most are present in several types.

Coastal dunes

Mobile, unconsolidated sands, exposed to salt-laden winds from the ocean. Plants are low and spreading, often with extensive root systems.

Typical species: *Spinifex hirsutus, S. longifolius, Isolepis nodosa, Lepidosperma gladiatum, Carpobrotus virescens, Tetragonia decumbens, Scaevola crassifolia, Olearia axillaris,* and *Leucophyta brownii*

Coastal heath

Dense shrubland varying in height from a few centimetres to 2 m or more. Vegetation tends to be pruned by the wind to a uniform level and is therefore usually thicker and taller in gullies and other sheltered situations, sparser on the ridge-tops.

Typical species: *Exocarpos sparteus, Acacia littorea, Dryandra sessilis, Isotropis cuneifolia, Templetonia retusa, Boronia alata, Diplolaena dampieri, Spyridium globulosum, Agonis flexuosa, Melaleuca systena, Pimelea ferruginea, Thomasia triphylla, Anthocercis littorea, Leucopogon parviflorus, Scaevola nitida*

Peppermint woodland

Dense or open woodland to 4 or 5 m, with an understorey of shrubs and herbs, found on sandy limestone soils along sheltered parts of the Leeuwin–Naturaliste Ridge. Old peppermint trees may be gnarled and contorted into fascinating shapes.

Typical species: *Lepidosperma gladiatum, L. squamatum, Sowerbaea laxiflora, Conostylis aculeata, Phyllanthus calycinus, Hardenbergia comptoniana, Hibbertia furfuracea, Philotheca spicata, Hypocalymma robustum, Hybanthus calycinus, Brachyscome iberidifolia, Craspedia variabilis*

Banksia woodland

Either *Banksia attenuata* or *B. ilicifolia* may be the dominant species on deep sands behind the Leeuwin–Naturaliste Ridge. These trees are usually scattered, with an understorey of mostly medium and low shrubs, herbs and sedges.

Typical species: *Anarthria prolifera, Dasypogon bromeliifolius, Gompholobium capitatum, Jacksonia horrida, Pultenaea reticulata, Kunzea recurva, Melaleuca thymoides, Andersonia caerulea, Lysinema ciliatum,* and *Leucopogon cordatus*

Sheoak woodland

Common sheoak, *Allocasuarina fraseriana,* is sometimes the dominant tree, or more often mixed with Jarrah or Banksia on thin, sandy soils over granite. The sheoak's needle-like leaves cover the ground and discourage seedlings, so the understorey is usually fairly open. Sheoak woodlands are often a good place to find orchids.

Typical species: *Anarthria scabra, Adenanthos meisneri, Hovea pungens,* and *Kunzea* spp.

Jarrah–Marri forest

Jarrah, *Eucalyptus marginata,* and Marri, *Corymbia calophylla,* grow together in varying proportions, with Jarrah tending to dominate on thin lateritic, ironstone soils, and on bare, rocky areas. On sandy soils the understorey is denser and Marri is often more common. Factors such as the burning regime, detailed soil profile, slope and other site factors make for a huge variety of species combinations within the Jarrah–Marri forest.

Typical species: *Podocarpus drouynianus, Macrozamia riedlei, Xanthorrhoea preissii, Patersonia umbrosa* var. *xanthina, Grevillea quercifolia, Hakea lissocarpha, Banksia grandis, Persoonia longifolia, Adenanthos barbiger, Tetratheca setigera, Stackhousia monogyna, Daviesia decurrens, Bossiaea ornata, B. linophylla, Hibbertia hypericoides, Hypocalymma robustum, Pimelea spectabilis, Lechenaultia biloba,* and *Dampiera linearis*

Karri forest

We have a small outlier of Karri forest here. It runs from Boranup south to Augusta on sandy loam soils over limestone on the sheltered eastern side of the Leeuwin–Naturaliste Ridge. There are further isolated patches north of the main belt, mostly in the sheltered valleys of the Margaret River and other streams and this type of forest would certainly have been more extensive before land was cleared for agriculture. However, it has been cut off from the main Karri forest area from Pemberton through to the south coast and some of its associated understorey species are unique to this area.

Typical species: *Clematis pubescens, Paraserianthes lophantha, Acacia subracemosa, A. alata, Bossiaea disticha, Hovea elliptica, Hibbertia cuneiformis, Chorilaena quercifolia, Trymalium floribundum, Agonis flexuosa,* and *Pimelea sylvestris*

Sandplains

Extensive sandflats occur mainly just to the east of the Leeuwin–Naturaliste Ridge and behind the south coast sand-dunes. They generally have sandy or peaty soils over clay and may be damp, though not waterlogged, in winter. They usually support low heathland vegetation with occasional taller shrubs or trees. In some drier areas on the limestone Balga (*Xanthorrhoea preissii*) is the dominant species, with an understorey of low shrubs and herbs.

Typical species: *Anarthria scabra, Evandra aristata, Xanthorrhoea preissii, Dasypogon bromeliifolius, Patersonia occidentalis, Phlebocarya ciliata, Banksia littoralis, Adenanthos obovatus, Drosera* spp., *Astartea* spp., *Beaufortia sparsa, Kunzea* spp., and *Pimelea longiflora*

Winter-wet swamps and wetlands

A little lower than the sandplains and inundated in winter, these areas are harder to get to, but can be fascinating for plant enthusiasts! Different species tend to dominate different wetland areas depending on small variations in height that determine the amount of waterlogging each receives. They usually contain many plants of the Restionaceae family that are particularly difficult to identify.

Typical species: *Leptocarpus* spp., *Meeboldina* spp., *Lepidosperma tetraquetrum, Xyris* spp., *Hakea linearis, Eutaxia virgata, Bossiaea rufa, Agonis linearifolia, A. parviceps, Homalospermum firmum, Melaleuca lateritia,* and *M. preissiana*

Watercourses and stream banks (bottom right)

Similar to the last, but with running water. There is often a narrow line of dense vegetation along the banks of flowing waterways.

Typical species: mosses and ferns, *Baumea* spp., *Lepidosperma effusum, Gahnia decomposita, Callistachys lanceolata, Agonis linearifolia,* and *Boronia fastigiata*

Granite outcrops

Granite outcrops are always very special places with a distinctive flora of interesting little plants finding scarce footholds in rock crevices and in moist soil pockets amongst the moss swards. The plants here have to be tough enough to withstand extremes of both temperature and moisture.

Typical species: mosses and ferns, *Stypandra glauca, Borya scirpoidea, Hakea trifurcata, Drosera* spp., *Dillwynia* sp. *A Perth Flora (R. Coveny 8036), Darwinia citriodora, Thryptomene saxicola, Dodonaea ceratocarpa, Utricularia multifida,* and *Stylidium* spp.

Glossary

We have tried to keep the technical terms in this book to a minimum. The following definitions aim to clarify the few that we needed to include.

Plant nomenclature

All except two plants in this book belong to the major division of the plant world known as the **Angiospermae**, or flowering plants, in which the seeds of the plant are borne in an ovary. The other two, *Macrozamia riedlei* and *Podocarpus drouynianus*, belong to the division of the **Gymnospermae**, in which the seeds are borne not within an ovary, but naked on cone scales.

The main grouping of plants within these divisions is the **family**, the names of which always end in **-aceae**; for example Proteaceae, Droseraceae and Asteraceae. Within each family, plants sharing similar characteristics are grouped according to their **genus** (*pl.* **genera**), within which they are divided into **species** (*pl.* **species**, abbreviated to **sp**. and **spp**. respectively)

Every plant has a scientific name that has two parts, giving its genus and its species; for example: *Leschenaultia* (genus) *biloba* (species); *Eucalyptus* (genus) *marginata* (species).

In some instances, where it varies significantly over its range, a species may be further divided into **sub-species** (subsp. *pl.* subspp.), or **varieties** (var. *pl.* vars.). Where a formal name has not yet been assigned to a plant it is given a **phrase name**. This indicates its rank (*eg.* sp., subsp., var.) and then gives a geographic or morphological identifier and the collector's name and number representing a particular herbarium specimen; for example: *Astartea* sp. *Scott River (D. Backshall 88233)*. The process of describing, naming and publishing a new species may take several years. Until a new name is formalised by publication, its informal status is indicated by the letters 'ms' = **manuscript name**.

Parts of the plant

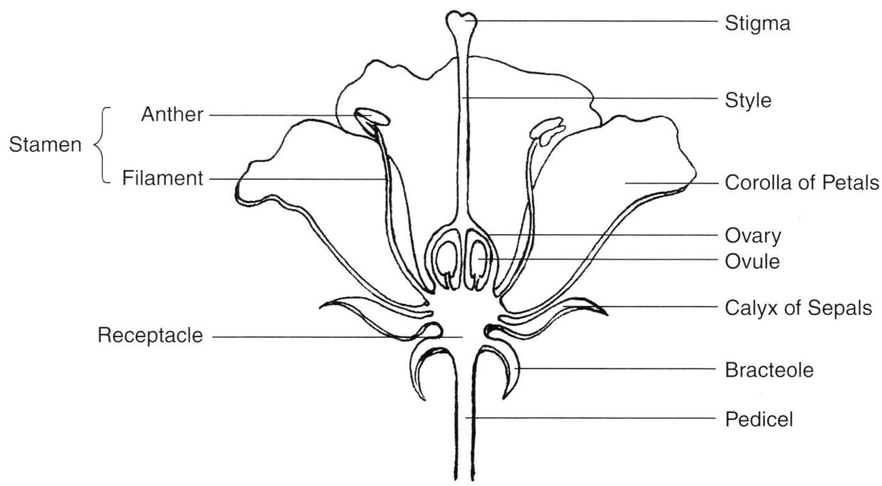

A flower consists of several parts arranged, regularly or irregularly, in concentric rings (see diagram). Note: Not all of these parts are necessarily present. The outer ring is the **calyx**, usually made up of distinct **sepals**. Inside this is the **corolla**, made up of **petals** that surround the plant's male organ, the **stamen**, usually several or many. The part of the stamen that bears the pollen is the **anther**, borne on a stalk called the **filament**. In the centre of the flower is the female organ, consisting of the **ovary**, bearing **ovules** that become **seeds** once they are fertilised, above which is the **style**, usually elongated, tipped by the **stigma** that receives the pollen.

Some species have separate male and female flowers and these may be **monoecious**, with both sexes on the same plant, or **dioecious**, with each sex on different plants.

Small leaf-like structures attached singly or in pairs to the calyx of a single flower are called **bracteoles**. **Bracts** are larger and are associated with the inflorescence as a whole (see below). These may look like petals, notably in many of the genus *Darwinia*. **Involucral bracts** are particularly large bracts, often borne in a whorl around the inflorescence.

The **inflorescence** refers to the arrangement of a group of flowers on a stem.

Individual flowers may be arranged on **spikes** of sessile flowers; in **racemes**, with a single axis of pedicellate flowers; in **panicles** that are compound racemes; **corymbs**, in which all the flowers are grouped in a horizontal plane; or **umbels**, in which all the pedicels arise from the same point (see illustrations).

A **pedicel** is an individual flower stalk.

A **peduncle** is the stalk of an inflorescence.

The **receptacle** is the swollen end of the pedicel that bears one or more flowers. In the Asteraceae (daisy family), this is particularly large.

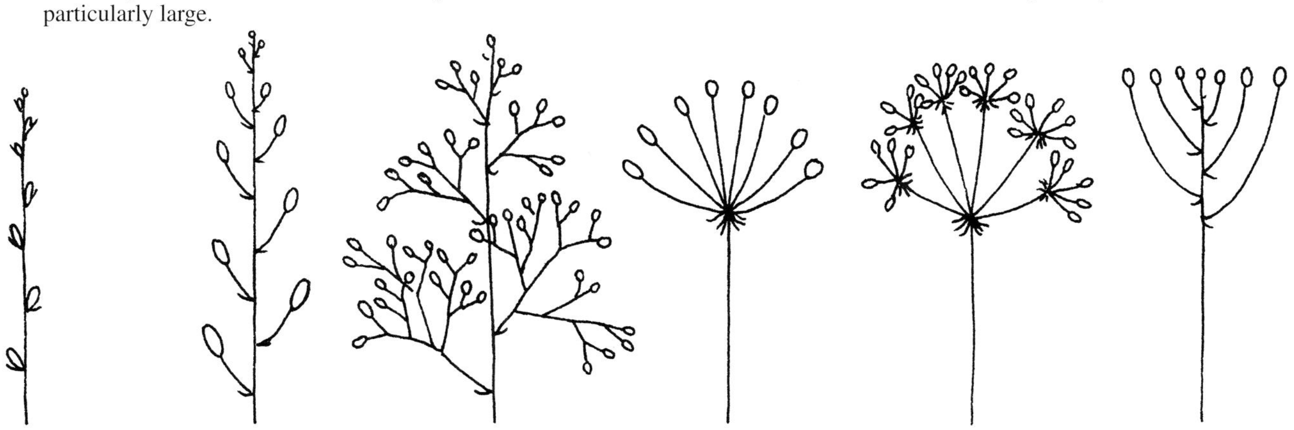

Spike Raceme Panicle Umbel Umbel Corymb

Phyllodes are enlarged, flattened pedicels (leaf-stalks) that have taken over the photosynthetic function normally performed by leaves. They are particularly apparent in many *Acacia* and *Daviesia* species.
Rhizomes are creeping stems, usually below ground, with roots arising from their nodes.
Tubers are swollen underground stems used as food storage organs. Roots may also be tuberous.

Plant habit

Plants are generally referred to as:
trees – over 5 m, with a single, woody trunk;
mallees – low trees or tall shrubs with a growth habit having several distinct, woody trunks arising separately from an underground rootstock known as a lignotuber – generally used in connection with eucalypts;
shrubs – less than 5 m, with several woody branches and without a main stem except maybe at ground level;
undershrubs (subshrubs) – woody, but less than 1 m high;
herbs – non-woody, or only woody at the base. Herbs may be:
annual – completing their full growth cycle to fruiting within a year, then dying
biennial – completing their full growth cycle over more than one, but within two years
perennial – with a life-span extending over more than two growing seasons
Endemism (endemic) is a term simply meaning that the plant species, or genus, or family, has a natural distribution confined to a particular geographic region; for example Western Australia, or the south-west of WA.

Bibliography

Bennett, E.M. & Dundas, P.J. 1988. *The Bushland Plants of Kings Park, Western Australia*, Kings Park and Botanic Garden, Perth.
Blackall, W.E. & Grieve, B.J. 1974 - 1998. *How to Know W.A. Wildflowers (Vols. 1 - 4),* University of Western Australia Press, Perth.
Corrick, M.G., Fuhrer, B.A. & George, A.S. 1996. *Wildflowers of Southern W.A.*, Five Mile Press, Monash University, Victoria.
Erickson, R., George, A.S., Marchant, N.G. & Morcombe, M.K. 1973. *Flowers & Plants of Western Australia*, Reed Books, Sydney.
French, M.E. & Nicolle, D. 1997. *Special Eucalypts of Perth & the South West*, F & N Publications, Perth.
Hoffman, N. & Brown, A. 1992. *Orchids of South-West Australia*, University of Western Australia Press, Perth.
Hopper, S.D. 1993. *Kangaroo Paws & Catspaws: a Natural History & Field Guide*, Department of Conservation and Land Management, Perth.
Hussey, B.M.J., Keighery, G.J., Cousens, R.D., Dodd, J. & Lloyd, S.G. 1997. *Western Weeds: a Guide to the Weeds of Western Australia*, Plant Protection Society of Western Australia, Victoria Park.
Marshall, J. & members of the Darling Range Branch, Wildflower Society of Western Australia 1996. *Wildflowers of the Western Hills*, Quality Publishing Australia.
Meney, K.A. & Pate, J.S. 1999. *Australian Rushes*, University of Western Australia Press, Perth.
Nevill, S. 1998. *A Guide to the Wildflowers of South Western Australia*, Simon Nevill Publications, Perth.
Powell, R. 1990. *Leaf & Branch: Trees and Tall Shrubs of Perth*, Department of Conservation and Land Management, Perth.
Paczkowska, G. & Chapman, A.R. 2000. *The Western Australian Flora: a Descriptive Catalogue*, Wildflower Society of Western Australia, Western Australian Herbarium, Department of Conservation and Land Management & the Botanic Gardens & Parks Authority.
Rippey, E. & Rowland, B. 1995. *Plants of the Perth Coast & Islands*, University of Western Australia Press, Perth.
Sharr, F.A. 1988. *Western Australian Plant Names & their Meanings: a Glossary,* University of Western Australia Press, Perth.
Young, J. 2000. *Hakeas of Western Australia: Irwin & Darling.*
Also:
Department of Conservation and Land Management Bushbook Series - a growing number of titles, including several little books on plants of the south-west.
and
Western Australian Herbarium 1998. *FloraBase - Information on the Western Australian flora*, Department of Conservation and Land Management. http://www.calm.wa.gov.au/science/florabase.html

Appendix – Species Checklist

You may photocopy this list if you wish to keep a record of species that you find in a particular area such as your block, a particular bushland reserve, along a walk-trail or road verge, or total species that you have identified.

Note that the order of species in the list is the same as their order in this book.

Area surveyed: ....

Date(s) of survey: From .... **To** ....

Species	Record ✓	Flowering Date/ Month
Macrozamia riedlei		
Podocarpus drouynianus		
Spinifex hirsutus		
Spinifex longifolius		
**Ammophila arenaria*		
Anarthria prolifera		
Anarthria scabra		
Lyginia imberbis		
Meeboldina scariosa		
Loxocarya cinerea		
Hypolaena exsulca		
Hypolaena pubescens		
Lepidosperma effusum		
Lepidosperma tetraquetrum		
Lepidosperma squamatum		
Lepidosperma gladiatum		
Lepidosperma pubisquamatum		
Tetraria octandra		
Cyathochaeta avenacea		
Gahnia decomposita		
Gahnia trifida		
Evandra aristata		
Schoenus grandiflorus		
Mesomelaena tetragona		
Isolepis nodosa		
Baumea vaginalis		
Juncus kraussii subsp. *australiensis*		
**Juncus microcephalus*		
Juncus pallidus		
**Typha orientalis*		
Xanthorrhoea brunonis subsp. *semibarbata*		
Xanthorrhoea preissii		
Xanthorrhoea gracilis		
Kingia australis		
Dasypogon hookeri		

Species	Record ✓	Flowering Date/ Month
Calectasia grandiflora subsp. *'southern' (H. Demarz 546)*		
Dasypogon bromeliifolius		
Lomandra nigricans		
Lomandra purpurea		
Lomandra caespitosa		
Acanthocarpos preissii		
Caesia micrantha		
Hodgsoniola junciformis		
Johnsonia lupulina		
Sowerbaea laxiflora		
Tricoryne elatior		
Tricoryne humilis		
Thysanotus arenarius		
Thysanotus dichotomus		
Thysanotus manglesianus		
Thysanotus multiflorus		
Thysanotus tenellus		
Agrostocrinum stypandroides		
Chamaescilla corymbosa var. *corymbosa*		
Dianella revoluta		
Stypandra glauca		
Burchardia multiflora		
Burchardia umbellata		
Wurmbea dioica subsp. *alba*		
Borya scirpoidea		
Patersonia occidentalis		
Patersonia umbrosa var. *xanthina*		
Orthrosanthus polystachyus		
Xyris roycei		
Philydrella pygmaea		
Hypoxis gardneri		
Hypoxis occidentalis var. *quadriloba*		
Anigozanthos manglesii		
Anigozanthos flavidus		

Species	Record ✓	Flowering Date/ Month
Anigozanthos viridus		
Conostylis aculeata subsp. *aculeata*		
Conostylis laxiflora		
Conostylis setigera subsp. *setigera*		
Phlebocarya ciliata		
Tribonanthes australis		
Haemodorum laxum		
Haemodorum simplex		
Haemodorum spicatum		
Thelymitra fuscolutea		
Thelymitra aff. *macrophylla*		
Elythranthera brunonis		
Prasophyllum brownii		
Cryptostylis ovata		
Praecoxanthus aphyllus		
Leporella fimbriata		
Eriochilus dilatatus		
Prasophyllum parvifolium		
Pterostylis rogersii		
Pterostylis aff. *nana*		
Pterostylis vittata		
Cyanicula deformis		
Corybas recurvus		
Diuris aff. *amplissima*		
Pterostylis recurva		
Pterostylis barbata		
Lyperanthus serratus		
Cyrtostylis huegelii		
Drakaea glyptodon		
Paracaleana nigrita		
Caladenia applanata		
Caladenia cairnsiana		
Caladenia citrina		
Caladenia ferruginea		
Caladenia flava		
Caladenia macrostylis		
Cyanicula sericea		
Allocasuarina fraseriana		
Allocasuarina humilis		
Banksia grandis		
Banksia attenuata		
Banksia littoralis		
Banksia ilicifolia		
Dryandra sessilis var. *cordata*		

Species	Record ✓	Flowering Date/ Month
Dryandra lindleyana subsp. *lindleyana*		
Grevillea bronwenae		
Grevillea quercifolia		
Grevillea pulchella subsp. *ascendens*		
Grevillea trifida		
Grevillea manglesioides subsp. *manglesioides*		
Hakea amplexicaulis		
Hakea prostrata		
Hakea lissocarpha		
Hakea trifurcata		
Hakea lasianthoides		
Hakea oleifolia		
Hakea linearis		
Hakea ruscifolia		
Hakea ceratophylla		
Adenanthos barbiger subsp. *intermedius* ms		
Adenanthos meisneri		
Adenanthos obovatus		
Synaphea gracillima		
Synaphea sp. *Redgate Rd. (J. Scott 16)*		
Xylomelon occidentale		
Persoonia elliptica		
Persoonia longifolia		
Isopogon sphaerocephalus		
Petrophile linearis		
Petrophile serruiae		
Petrophile diversifolia		
Conospermum paniculatum		
Conospermum capitatum subsp. *glabratum*		
Nuytsia floribunda		
Cassytha racemosa		
**Orobanche minor*		
Exocarpos sparteus		
Leptomeria squarrulosa		
Santalum acuminatum		
Muehlenbeckia adpressa		
Rhagodia baccata subsp. *baccata*		
Atriplex isatidea		
Threlkeldia diffusa		
Sarcocornia blackiana		
Calandrinia liniflora		

Species	Record ✓	Flowering Date/ Month
Carpobrotus virescens		
**Tetragonia decumbens*		
Tetragonia implexicoma		
**Cakile maritima*		
Stenopetalum robustum		
Ptilotus manglesii		
Ptilotus drummondii		
Ranunculus colonorum		
**Ranunculus muricatus*		
Clematis pubescens		
Cheiranthera preissiana var. *planifolia*		
Sollya heterophylla		
Marianthus candidus		
Marianthus tenuis		
Billardiera floribunda		
Billardiera laxiflora		
Billardiera variifolia		
Hibbertia cuneiformis		
Hibbertia grossulariifolia		
Hibbertia furfuracea		
Hibbertia hypericoides		
Hibbertia cunninghamii		
Hibbertia inconspicua		
Hibbertia pilosa		
Hibbertia racemosa		
Hibbertia notibractea ms		
Hibbertia rhadinopoda		
Frankenia pauciflora		
Drosera glanduligera		
Drosera macrantha subsp. *macrantha*		
Drosera marchantii subsp. *marchantii*		
Drosera pallida		
Drosera erythrorhiza		
Drosera menziesii subsp. *menziesii*		
Drosera sulphurea		
Drosera stelliflora		
Drosera pulchella		
Utricularia multifida		
Paraserianthes lophantha var. *lophantha*		
Acacia cyclops		
Acacia cochlearis		
Acacia myrtifolia		

Species	Record ✓	Flowering Date/ Month
Acacia rostellifera		
Acacia saligna		
Acacia alata var. *alata*		
Acacia stenoptera		
Acacia obovata		
Acacia urophylla		
Acacia scalpelliformis		
Acacia extensa		
Acacia divergens		
Acacia mooreana		
Acacia littorea		
Acacia uliginosa		
Acacia gilbertii		
Acacia pulchella var. *pulchella*		
Acacia browniana var. *browniana*		
Acacia varia var. *varia*		
Acacia subracemosa		
Acacia aff. *lateriticola*		
Mirbelia dilatata		
Callistachys lanceolata		
Oxylobium lineare		
Gastrolobium bilobum		
Isotropis cuneifolia subsp. *cuneifolia*		
Chorizema aciculare subsp. *aciculare*		
Chorizema diversifolium		
Chorizema nanum		
Chorizema reticulatum		
Chorizema rhombeum		
Gompholobium capitatum		
Gompholobium confertum		
Gompholobium knightianum		
Gompholobium ovatum		
Gompholobium polymorphum		
Gompholobium preissii		
Gompholobium tomentosum		
Jacksonia horrida		
Viminaria juncea		
Sphaerolobium drummondii		
Sphaerolobium fornicatum		
Sphaerolobium medium		
Daviesia cordata		
Daviesia decurrens		
Daviesia horrida		

Species	Record ✓	Flowering Date/ Month
Daviesia inflata		
Daviesia preissii		
Pultenaea drummondii		
Pultenaea reticulata		
Eutaxia parvifolia		
Eutaxia obovata		
Eutaxia virgata		
Dillwynia sp. *A Perth Flora (R. Coveny 8036)*		
Hovea chorizemifolia		
Hovea elliptica		
Hovea pungens		
Hovea trisperma		
Hardenbergia comptoniana		
Bossiaea aquifolium subsp. *aquifolium*		
Bossiaea disticha		
Bossiaea linophylla		
Bossiaea ornata		
Bossiaea praetermissa		
Bossiaea rufa		
Templetonia retusa		
Kennedia carinata		
Kennedia coccinea		
Kennedia macrophylla		
Kennedia prostrata		
**Pelargonium capitatum*		
Pelargonium australe subsp. *drummondii*		
Pelargonium littorale		
Geranium solanderi		
Nitraria billardierei		
Beyeria viscosa		
Ricinocarpos glaucus		
Poranthera huegelii		
Phyllanthus calycinus		
Boronia alata		
Boronia anceps		
Boronia crenulata subsp. *crenulata*		
Boronia fastigiata subsp. *tenuior*		
Boronia gracilipes		
Boronia megastigma		
Boronia molloyae		
Diplolaena dampieri		
Chorilaena quercifolia		
Philotheca spicata		

Species	Record ✓	Flowering Date/ Month
Rhadinothamnus anceps		
Tremandra stelligera		
Tremandra diffusa		
Tetratheca hirsuta		
Tetratheca setigera		
Comesperma calymega		
Comesperma ciliatum		
Comesperma confertum		
Comesperma flavum		
Dodonaea aptera		
Dodonaea ceratocarpa		
Stackhousia monogyna		
Tripterococcus brunonis		
Trymalium floribundum subsp. *trifidum*		
Trymalium ledifolium var. *rosmarinifolium*		
Spyridium globulosum		
Cryptandra arbutiflora var. *tubulosa*		
Astartea sp. *Scott River (D. Backshall 88233)*		
Astartea sp. *juniperina (G.J. Keighery 9558)*		
Hypocalymma angustifolium		
Hypocalymma robustum		
Calytrix flavescens		
Agonis flexuosa var. *flexuosa*		
Agonis linearifolia		
Agonis parviceps		
Agonis juniperina		
Homalospermum firmum		
Darwinia citriodora		
Darwinia thymoides subsp. *thymoides*		
Verticordia plumosa var. *plumosa*		
Thryptomene saxicola		
Pericalymma ellipticum var. *ellipticum*		
Kunzea ciliata		
Kunzea recurva		
Kunzea spathulata		
Beaufortia sparsa		
Calothamnus sanguineus		
Calothamnus lateralis		
Melaleuca systena		
Melaleuca incana subsp. *incana*		
Melaleuca thymoides		
Melaleuca lateritia		

Species	Record ✓	Flowering Date/ Month
Melaleuca basicephala		
Melaleuca huegelii subsp. *huegelii*		
Melaleuca lanceolata		
Melaleuca preissiana		
Melaleuca rhaphiophylla		
Melaleuca cuticularis		
Eucalyptus marginata		
Eucalyptus patens		
Eucalyptus diversicolor		
Eucalyptus megacarpa		
Eucalyptus cornuta		
Eucalyptus calcicola		
Corymbia calophylla		
Thomasia foliosa		
Thomasia grandiflora		
Thomasia macrocarpa		
Thomasia paniculata		
Thomasia triphylla		
Lasiopetalum floribundum		
Pimelea clavata		
Pimelea hispida		
Pimelea longiflora subsp. *longiflora*		
Pimelea suaveolens subsp. *suaveolens*		
Pimelea ferruginea		
Pimelea rosea subsp. *rosea*		
Pimelea spectabilis		
Pimelea sylvestris		
Pimelea angustifolia		
Acrotriche cordata		
Astroloma ciliatum		
Astroloma pallidum		
Astroloma sp. Nannup *(R.D. Royce 3978)*		
Conostephium pendulum		
Styphelia tenuiflora		
Leucopogon australis		
Leucopogon capitellatus		
Leucopogon parviflorus		
Leucopogon revolutus		
Leucopogon verticillatus		
Leucopogon assimilis		
Leucopogon cordatus		
Leucopogon elatior		
Leucopogon propinquus		
Leucopogon sp. Margaret River *(J. Scott 207)*		

Species	Record ✓	Flowering Date/ Month
Leucopogon pendulus		
Andersonia caerulea		
Andersonia involucrata		
Lysinema ciliatum		
Lysinema conspicuum		
Sphenotoma capitatum		
Sphenotoma gracile		
Samolus repens		
Logania serpyllifolia subsp. *angustifolia*		
Logania vaginalis		
Hybanthus calycinus		
Hemiandra australis		
Hemigenia rigida		
Hemigenia podalyrina		
Platysace filiformis		
Platysace tenuissima		
Trachymene coerulea subsp. *coerulea*		
Trachymene pilosa		
Xanthosia candida		
Xanthosia tasmanica		
Xanthosia atkinsoniana		
Pentapeltis peltigera		
Apium prostratum var. *prostratum*		
Eryngium pinnatifidum subsp. *pinnatifidium*		
Centella asiatica		
Dichondra repens		
Villarsia latifolia		
Villarsia parnassifolia		
Opercularia echinocephala		
Opercularia hispidula		
Opercularia vaginata		
Opercularia volubilis		
Solanum symonii		
Anthocercis littorea		
Myoporum insulare		
Myoporum oppositifolium		
Veronica distans		
Lobelia alata		
Lobelia gibbosa		
Lobelia tenuior		
Isotoma hypocrateriformis		
Isotoma scapigera		
Wahlenbergia littoricola		

Species	Record ✓	Flowering Date/ Month
Goodenia eatoniana		
Goodenia pusilla		
Velleia trinervis		
Lechenaultia biloba		
Dampiera alata		
Dampiera hederacea		
Dampiera linearis		
Dampiera trigona		
Scaevola calliptera		
Scaevola crassifolia		
Scaevola glandulifera		
Scaevola microphylla		
Scaevola nitida		
Diaspasis filifolia		
Levenhookia pusilla		
Levenhookia stipitata		
Stylidium amoenum		
Stylidium calcaratum		
Stylidium piliferum		
Stylidium petiolare		
Stylidium schoenoides		
Stylidium spathulatum subsp. *spathulatum*		
Stylidium caespitosum		
Stylidium fasciculatum		
Stylidium junceum subsp. *junceum*		
Stylidium laciniatum		
Stylidium violaceum		
Stylidium crassifolium		
Stylidium dichotomum		
Stylidium eriopodum		
Trichocline spathulata		
Olearia axillaris		
Olearia elaeophila		
Olearia paucidentata		
Olearia rudis		
Asteridea pulverulenta		
Brachyscome iberidifolia		
Rhodanthe citrina		
Leucophyta brownii		
Siloxerus humifusus		
Senecio lautus subsp. *maritimus*		
Senecio ramossissimus		
Hyalosperma cotula		
Podolepis gracilis		

Species	Record ✓	Flowering Date/ Month
Podolepis lessonii		
Craspedia variabilis		
Ozothamnus cordatus		
Ozothamnus ramosus		
Lagenophora huegelii		
**Cotula turbinata*		
**Amaryllis belladonna*		
**Zantedeschia aethiopica*		
**Trachyandra divaricata*		
**Gladiolus undulatus*		
**Moraea (Homeria) flaccida*		
**Ferraria crispa*		
**Sparaxis bulbifera*		
**Romulea rosea*		
**Ixia* aff. *maculata*		
**Chasmanthe floribunda*		
**Watsonia meriana* var. *bulbillifera*		
**Watsonia versfeldii*		
**Asparagus asparagoides*		
**Oenothera glazioviana*		
**Oxalis pes-caprae*		
**Oxalis purpurea*		
**Vinca major*		
**Anagallis arvensis*		
**Leptospermum laevigatum*		
Leptospermum erubescens		
**Pittosporum undulatum*		
**Dipogon lignosus*		
**Genista monspessulana*		
**Podalyria sericea*		
**Chamaecytisus palmensis*		
**Centaurium tenuiflorum*		
**Bartsia trixago*		
**Petrorhagia dubia*		
**Arctotheca calendula*		
**Arctotheca populifolia*		
**Senecio elegans*		
**Solanum linneanum*		
**Emex australis*		
**Fumaria capreolata*		
**Fumaria muralis*		
**Silene gallica* var. *gallica*		
**Euphorbia paralias*		
Echium plantagineum		
**Verbascum virgatum*		

Index

Jane Scott

Jane has an honours degree in geography from University College London, and qualifications in teaching and librarianship, but her passions are wildflowers and bushwalking. Western Australia's unique wildflowers have fascinated her ever since she arrived in Australia from England in 1973. During 25 years of living and working in Margaret River Jane has developed a wide knowledge of the area's plants and their habitats, and the production of a book on local flora has long been a dream. The realisation of this ambition owes much to her close working partnership with artist, Patricia Negus, and extensive professional help from staff of the Western Australian Herbarium.

Patricia Negus

Patricia was born and educated in England, obtaining an honours degree in biology and diploma of education from Reading University. She worked as a biology teacher for several years, before becoming a full-time artist in 1976. Patricia has exhibited widely in the south-west of Western Australia, winning many awards. She has a particular love of nature, and an eye for detail. This, her third book, has been a great challenge, and has given much satisfaction in seeing its completion.